T0235779

SpringerBriefs in Applied Sciences and Technology

SpringerBriefs in Mathematical Methods

Series Editors

Anna Marciniak-Czochra, Institute of Applied Mathematics, IWR, University of Heidelberg, Heidelberg, Germany

Thomas Reichelt, Emmy-Noether Research Group, Universität Heidelberg, Heidelberg, Germany

More information about this subseries at https://link.springer.com/bookseries/11219

Jian Li · Xiaolin Lin · Zhangxing Chen

Finite Volume Methods for the Incompressible Navier–Stokes Equations

 Springer

Jian Li
Shaanxi University of Science
and Technology
Xi'an, China

Xiaolin Lin
Shaanxi University of Science
and Technology
Xi'an, China

Zhangxing Chen
University of Calgary
Calgary, AB, Canada

ISSN 2191-530X   ISSN 2191-5318 (electronic)
SpringerBriefs in Applied Sciences and Technology
ISSN 2365-0826   ISSN 2365-0834 (electronic)
SpringerBriefs in Mathematical Methods
ISBN 978-3-030-94635-7   ISBN 978-3-030-94636-4   (eBook)
https://doi.org/10.1007/978-3-030-94636-4

This Springer imprint is published by the registered company Springer Nature Switzerland AG
The registered company address is: Gewerbestrasse 11, 6330 Cham, Switzerland

To Megan and Jeremy.

Preface

Recently, engineers, scientists, and mathematicians have concentrated their efforts on numerical methods for solving the Navier–Stokes equations. The reason is that these equations are used in important scientific and industrial areas such as computational fluid dynamics, underground water flow, and multi-phase flow in petroleum reservoirs. While the finite difference and element methods have been extensively covered, few books have dealt with the finite volume methods for these equations. The latter methods have attracted much attention because of their special features in both handling complex geometries and achieving fast solution speed. On the other hand, their numerical analysis needs special attention since their trial and test functions differ.

The purpose of this book is to provide a fairly comprehensive understanding of the most recent developments on the finite volume methods. Its focus is the development and analysis of these methods for the two- and three-dimensional Navier–Stokes equations. It covers the lower order finite element pairs, well-posedness, and optimal analysis for these finite volume methods. In presenting the materials, we attempt to make this book self-contained by offering complete proofs and/or recalling theoretical results. While most of these materials have been taught by the authors in a number of institutions over the past several years, we also include several updated lemmas and theorems for the finite volume methods for the incompressible Navier–Stokes equations.

This book is primarily developed to address research needs for undergraduate and graduate students and academic and industrial researchers. It is particularly valuable as a research reference in the fields of engineering, mathematics, physics, and computer sciences.

We express our thanks to the staff at the Springer-Verlag Heidelberg office for their expert guidance and assistance with the edition of this book as well as their advice throughout the book preparation. Moreover, we are particularly grateful to the fund by NSF of China (Grant No. 11771259), special support program to develop innovative talents in the region of Shaanxi province, and innovation team on computationally efficient numerical methods based on new energy problems in Shaanxi province.

We gratefully thank Profs. Xiaobing Feng, Yinnian He, and Yanping Lin for their collaborations and constructive comments over the past several years. To our families, we dedicate this book to them in token of gratitude for their love, forbearance, and understanding.

Xi'an, China   Jian Li
Xi'an, China   Xiaolin Lin
Calgary, Canada   Zhangxin Chen
February 2021

Contents

1 **Mathematical Foundation** 1
  1.1 L^p Spaces 1
  1.2 Sobolev Spaces 2
  1.3 Imbedding Inequalities 4
  1.4 The Navier–Stokes Equations 4
    1.4.1 Mathematical Model 5
    1.4.2 Numerical Solution 6

2 **FVMs for the Stationary Stokes Equations** 9
  2.1 Introduction 9
  2.2 Weak Formulation 11
  2.3 Galerkin FV Approximation 13
  2.4 Existence and Uniqueness Theorems 19
  2.5 Priori Estimate 22
    2.5.1 Superclose 22
    2.5.2 Optimal Analysis 23
    2.5.3 L^2 Estimate for Velocity 24
    2.5.4 Optimal L^∞ Estimate 25
  2.6 Posteriori Estimation 32
    2.6.1 Upper Bound 32
    2.6.2 Lower Bound 35
  2.7 Adaptive Mixed Finite Volume Methods 38
    2.7.1 Discrete Local Lower Bound 39
    2.7.2 Adaptive Finite Volume Algorithms 42
    2.7.3 Convergence Analysis 44
  2.8 Numerical Experiments 47
  2.9 Conclusions 50

3 **FVMs for the Stationary Navier–Stokes Equations** 53
  3.1 Introduction 53
  3.2 FVMs for the Stationary Navier–Stokes with Small Data 54
    3.2.1 The Weak Formulation 54

  3.2.2   Galerkin FV Approximation 56
  3.2.3   Existence and Uniqueness Theorem 57
  3.2.4   Convergence Analysis 61
  3.2.5   Optimal L^∞ Analysis 66
3.3 FVEs of Branches of Nonsingular Solutions 74
  3.3.1   An Abstract Framework 74
  3.3.2   Existence and Uniqueness 76
  3.3.3   Optimal Analysis 77
  3.3.4   Optimal L^∞ Estimate 81
3.4 Numerical Experiments 81
3.5 Conclusions 82

4 FVMs for the Nonstationary Navier–Stokes Equations 85
4.1 Introduction 85
4.2 The Weak Formulation 86
4.3 Galerkin FV Approximation 89
4.4 Stability and Error Analysis 93
4.5 L^2-Error Estimates 108
4.6 Conclusions 113

References 115

Index 121

Chapter 1
Mathematical Foundation

Abstract The purpose of this chapter is to recall the basic concepts related to the Navier–Stokes (NS, in short) equations, including the Sobolev spaces, several important inequalities for analysis, and the NS equations. For more details, we refer the reader to the monographs [1, 59, 122] and references therein.

1.1 L^p Spaces

The L^p spaces are also called the Lebesgue spaces. They form an important class of Banach spaces in the functional analysis and topological vector spaces. The Lebesgue spaces have been widely applied in many fields. In this section, we give them a brief introduction.

Let $\Omega \subset \mathbb{R}^N$ be a non-null measurable set. A point in Ω is denoted by $x = (x_1, x_2, \ldots, x_N)$. For $1 \leq p < \infty$, we define the following space of equivalence classes of measurable functions:

$$L^p(\Omega) = \left\{ v : \Omega \to R \text{ such that } \int_\Omega |v|^p dx < \infty \right\},$$

with the norm

$$\|v\|_{L^p(\Omega)} = \left(\int_\Omega |v|^p dx \right)^{1/p}.$$

For $p = \infty$, we define $L^\infty(\Omega)$ to be the space of functions which are bounded a.e. in Ω, with the norm

$$\|v\|_{L^\infty(\Omega)} = \operatorname*{ess\,sup}_{x \in \Omega} |v(x)|$$
$$= \inf\{C \geq 0 : |v(x)| \leq C \text{ a.e. in } \Omega\}$$
$$= \sup\{|v(x)| \text{ a.e. in } \Omega\}.$$

J. Li et al., *Finite Volume Methods for the Incompressible Navier–Stokes Equations*, SpringerBriefs in Mathematical Methods, https://doi.org/10.1007/978-3-030-94636-4_1

For the equivalence classes in the L^p spaces, two functions are identical if they are equal almost everywhere; i.e., they differ at most on a subset of Ω with zero measure. Furthermore, we define a locally integrable function to be a function that is integrable on every compact subset of its domain of definition. The importance of such functions lies in the fact that their function space is similar to the L^p spaces, but its members are not required to satisfy any growth restriction on their behavior at infinity. In other words, locally integrable functions can grow arbitrarily fast at infinity, but they are still manageable in a way similar to ordinary integrable functions.

We recall two properties of the L^p spaces.

Property 1. The $L^p(\Omega)$, $1 \leq p \leq \infty$, spaces are Banach spaces.
Property 2. The $L^p(\Omega)$, $1 < p < \infty$, spaces are reflexive spaces.

Two useful inequalities are also recalled.

- **Hölder's inequality:** If $u \in L^p(\Omega)$ and $v \in L^q(\Omega)$, $\frac{1}{p} + \frac{1}{q} = 1$, then $uv \in L^1(\Omega)$ satisfying

$$\int_\Omega |uv| dx \leq \|u\|_{L^p} \|v\|_{L^q}, \tag{1.1}$$

where $q \in [1, \infty]$ is called the conjugate of p. In particular, if $p = q = 2$, it is the Cauchy-Schwarz inequality.

- **Generalized Hölder's inequality:** If $u \in L^p(\Omega)$, $v \in L^q(\Omega)$, and $w \in L^r(\Omega)$, $\frac{1}{p} + \frac{1}{q} + \frac{1}{r} = 1$, then $uvw \in L^1(\Omega)$ satisfying

$$\int_\Omega |uvw| dx \leq \|u\|_{L^p} \|v\|_{L^q} \|w\|_{L^r}. \tag{1.2}$$

In the context of this book, the spaces $L^p(\Omega)$, $p \in [1, \infty]$, and their norms are often abbreviated as L^p and $\| \cdot \|_{0,p}$ (or $\| \cdot \|_{L^p}$).

1.2   Sobolev Spaces

For a nonnegative integer k and a real number $q \in [1, \infty]$, the Sobolev spaces $W^{k,q}(\Omega)$ are the spaces of functions in $L^q(\Omega)$ whose distributional derivatives of order less than or equal to k belong to $L^q(\Omega)$. Let $\alpha = (\alpha_1, \alpha_2, \ldots, \alpha_N)$ and set $|\alpha| = \sum_{i=1}^N \alpha_i$. Denote

$$W^{k,q}(\Omega) = \{v : \|v\|_{k,q,\Omega} < \infty\},$$

with the norm

$$
\|v\|_{k,q,\Omega} =
\begin{cases}
\left(\sum\limits_{|\alpha| \le k} \int_\Omega \left| \frac{\partial^\alpha v(x)}{\partial x^\alpha} \right|^q dx \right)^{1/q} & \text{if } q < \infty, \\[4mm]
\max\limits_{|\alpha| \le k} \sup\limits_{x \in \Omega} \left| \frac{\partial^\alpha v(x)}{\partial x^\alpha} \right| & \text{if } q = \infty.
\end{cases}
$$

Let $W_0^{k,q}(\Omega)$ be the completion of $C_0^\infty(\Omega)$ according to the norm $\| \cdot \|_{k,q,\Omega}$, where $C_0^\infty(\Omega)$ represents the space of functions with continuous derivatives of arbitrary order and compact support on Ω.

For $1 \le p < \infty$, we define $W^{-k,p}(\Omega)$ to be the dual space of $W_0^{k,q}(\Omega)$, with the norm

$$
\|f\|_{-k,p,\Omega} = \sup_{0 \ne v \in W_0^{k,q}(\Omega)} \frac{<f, v>}{\|v\|_{k,q,\Omega}},
$$

where q is the conjugate of p and $< \cdot, \cdot >$ is the duality between $W_0^{k,q}(\Omega)$ and its dual.

In particular, if there is no ambiguity, the subscript $p = 2$ can be dropped when referring to the spaces $W^{k,2}(\Omega)$, with the following notation:

$$
H^k(\Omega) = W^{k,2}(\Omega), \ H_0^k(\Omega) = W_0^{k,2}(\Omega).
$$

They are Hilbert spaces with the inner product

$$
(u, v)_{k,\Omega} = \sum_{|\alpha| \le k} \int_\Omega \partial^\alpha u \partial^\alpha v \, dx \ \ \forall u, v \in H^k(\Omega)
$$

and the seminorm

$$
|v|_{k,\Omega} = \left(\sum_{|\alpha|=k} \int_\Omega |\partial^\alpha v|^2 dx \right)^{1/2}.
$$

For convenience, we will omit Ω and write $(\cdot, \cdot) = (\cdot, \cdot)_{0,\Omega}$, $\| \cdot \|_{k,p} = \| \cdot \|_{k,p,\Omega}$ and $\| \cdot \|_{-k,p} = \| \cdot \|_{-k,p,\Omega}$.

In order to deal with a time-dependent initial-boundary value problem, the corresponding definitions are introduced. Specifically, we consider space-time functions $v(x, t), x \in \Omega, t \in (0, T)$, whose functional space is introduced as follows:

$$
L^q([0, T]; W^{k,p}(\Omega))
$$
$$
= \left\{ v : [0, T] \to W^{k,p}(\Omega) \text{ such that } v \text{ is measurable and } \int_0^T \|v\|_{k,p}^q dt < \infty \right\},
$$

$$
(1.3)
$$

where $k \geq 0$, $1 \leq p \leq \infty$, $1 \leq q < \infty$, and v is measurable in both temporal and spatial scales, endowed with the norm

$$\|v\|_{L^q([0,T];W^{k,p}(\Omega))} = \left(\int_0^T \|v\|_{k,p}^q dt \right)^{1/q}.$$

Moreover, $C^0([0,T];W^{k,p}(\Omega))$ and $L^\infty([0,T];W^{k,p}(\Omega))$ can be defined in a similar way.

1.3 Imbedding Inequalities

In this section, we introduce several useful imbedding inequalities, which will be used frequently in the later chapters.

- Let $\Omega \subset \mathbb{R}^N$ be a domain. If $1 \leq p \leq q \leq \infty$, there holds:

$$L^q(\Omega) \subset L^p(\Omega) \subset L^1(\Omega) \subset L^1_{loc}(\Omega),$$

  where $L^1_{loc}(\Omega) = \{f : \Omega \to R \text{ such that } f|_{\mathscr{O}} \in L^1(\mathscr{O}) \text{ for each compact set } \mathscr{O} \subset \Omega\}$.
- Let $\Omega \subset \mathbb{R}^N$ be unbounded. There holds:

$$L^p(\Omega) \subset L^1_{loc}(\Omega) \quad \forall p \geq 1.$$

- If $n > 1$ and the boundary $\partial\Omega$ is Lipschitz continuous, there hold the following inclusions:

  – if $0 < 2s < n$, then $H^s(\Omega) \subset L^q(\Omega) \ \forall 1 \leq q \leq \frac{2n}{n-2s}$;
  – if $2s = n$, then $H^s(\Omega) \subset L^q(\Omega) \ \forall 1 \leq q < \infty$;
  – if $2s > n$, then $H^s(\Omega) \subset C^0(\bar{\Omega})$.

1.4 The Navier–Stokes Equations

Much phenomena in scientific and engineering fields can be modeled by the Navier–Stokes equations. For example, the Navier–Stokes equations in their complete and simplified forms are applied in the design of aircrafts and cars, the study of blood flow, the design of power stations, and the analysis of air pollution. Coupled with Maxwell's equations they can be used to model and study magnetohydrodynamics [137]. One poem says "Waves follow our boat as we meander across the lake, and turbulent air currents follow our flight in a modern jet". Mathematicians and physicists believe that an explanation for and the prediction of both the breeze and the turbulence can be found through an understanding of solutions to the Navier–Stokes equations. Although these equations were written down in the nineteenth century, our understanding of them remains minimal. The challenge is to make substantial progress toward a mathematical theory which will unlock the secrets hidden in the Navier–Stokes equations.

The Navier–Stokes equations are also of great interest in a purely mathematical sense. The Clay Mathematics Institute has called the existence and smoothness problem of the Navier–Stokes equations one of the seven most important open problems in mathematics and has offered a 1,000,000 US dollar prize for a solution or a counter-example.

1.4.1 Mathematical Model

The model describes the Navier–Stokes flow in a two- or three-dimensional fluid region Ω. Here Ω is a bounded domain and has a Lipschitz-continuous boundary or C^2 smooth boundary Γ.

The Navier–Stokes equations are stated as follows:

$$\rho u_t + \rho(u \cdot \nabla)u - \nabla \cdot \sigma(u, p) = \rho f \qquad \forall (x, t) \in \Omega \times (0, T], \qquad (1.4)$$

$$\nabla \cdot u = 0 \qquad \forall (x, t) \in \Omega \times (0, T], \qquad (1.5)$$

$$u(x, 0) = u_0(x) \qquad \forall x \in \Omega, \qquad (1.6)$$

$$u = 0 \qquad \forall (x, t) \in \Gamma \times (0, T]. \qquad (1.7)$$

Here, $u : \Omega \times [0, T] \to R^d$ and $p : \Omega \times [0, T] \to R$ represent velocity and pressure in Ω, respectively. $f : [0, T] \to [H^1(\Omega)]^d$ is the body force on per unit volume. Also, the deformation rate tensor $\mathbb{D}(u)$ and stress tensor $\sigma(u, p)$ associated with (u, p) are defined by $\mathbb{D}(u) = \frac{1}{2}[\nabla u + (\nabla u)^T]$ and $\sigma(u, p) = 2\mu\mathbb{D}(u) - p\mathbb{I}$, where $(\nabla u)^T$ denotes the transpose of the matrix ∇u and $\mathbb{I}$ is the identity tensor. In addition, $\rho > 0$ and $\mu > 0$ stand for the fluid density and viscosity, respectively.

In general, the viscosity coefficient depends on temperature. Here, we assume that the viscosity $\mu > 0$ is a positive constant. The density ρ also varies as a function of time and space. In the present context, we mainly concentrate on a viscous Newtonian fluid over a limited range of flow rates where turbulence, inertial, and other high-velocity effects are negligible.

Then, the incompressible Navier–Stokes equations can be simplified as follows:

$$u_t + (u \cdot \nabla)u - \mu\Delta u + \nabla p = f \qquad \forall (x, t) \in \Omega \times (0, T], \qquad (1.8)$$

$$\nabla \cdot u = 0 \qquad \forall (x, t) \in \Omega \times (0, T], \qquad (1.9)$$

$$u(x, 0) = u_0(x) \qquad \forall x \in \Omega, \qquad (1.10)$$

$$u = 0 \qquad \forall (x, t) \in \Gamma \times (0, T], \qquad (1.11)$$

which is a model describing a Newtonian, homogeneous, and incompressible fluid.

A canonical way is introduced by using the Reynolds number which measures the effect of viscosity on flow. Let L be a characteristic length and U a characteristic velocity. Then, the characteristic time is determined by $T = L/U$. Furthermore, we consider the non-dimensionalization by the transformations [25, 30, 59, 78, 102]:

$$\bar{u} = u/U, \, \bar{p} = p/U^2, \, \bar{x} = x/L, \, \bar{t} = t/T = T/LU,$$

which measure the effect of viscosity on flow.

By using these dimensionless quantities, the momentum equation (1.8) can be rewritten as follows:

$$\bar{u}_{\bar{t}} - (\bar{u} \cdot \nabla)\bar{u} - \frac{1}{Re}\Delta\bar{u} + \nabla\bar{p} = f \qquad \text{in } \Omega, \qquad (1.12)$$

where the Reynolds number is defined by $Re = LU/\mu$.

Flows with the same Reynolds number are similar in domains with the same shape. Therefore, one can construct an experiment using a practical size in the laboratory to model flows in large scales. The size of the smallest scales is set by the Reynolds number. As the Reynolds number increases, smaller and smaller scales of a flow are visible. The Reynolds number is an indicator of the range of scales in a flow. The higher the Reynolds number, the greater the range of scales. The largest eddies are always the same size; the smallest eddies are determined by the Reynolds number. From the definition of Re, for a large scale problem, its viscosity is tiny. In the limiting case with $Re = \infty$ or $\mu = 0$, the Navier–Stokes equations degrade into the so-called Euler equations, which describe an ideal fluid. Note that the Navier–Stokes equations are second order while the Euler equations are first order. If there is a mismatch in the boundary condition, it causes problems near the boundary, known as a boundary layer effect; see [30] for details. Here, we only concentrate on the laminar flow of low velocity, determined by the incompressible Navier–Stokes equations. The studies on turbulent flow are highly technical and beyond the scope of this book.

1.4.2 Numerical Solution

The numerical simulation of complex, dynamic processes that appear in nature or in industrial applications poses a lot of challenging mathematical problems, opening a long road from a basic problem to mathematical modeling, numerical simulation, and, finally, to the interpretation of simulation results.

In general, we cannot obtain a solution of the complex, dynamic processes in explicit form. Moreover, the available analytical integration methods are of limited applicability, which may result in the solution of mathematical problems that are generally rather involved because of the imposition of boundary conditions.

From a theoretical point of view, the analysis of the Navier–Stokes equations is often bound to investigating existence, uniqueness, and possible regularity of their solutions, which is relatively abundant in the 2D case and is still open in the 3D case. Therefore, it is extremely important to have numerical methods at one's disposal that allow to construct an approximation (u_N, p_N) of the exact solution (u, p) and to evaluate (in some suitable norms) the error

$$\||(u_N - u, p_N - p)\|| = (\|u_N - u\|_1 + \|p_N - p\|_0)^{1/2}.$$

For convenience, we construct an approximated problem $NS_N(u_N, p_N)$ to the original problem $NS(u, p)$. Then, we will recall some basic concepts of the numerical methods [108].

Definition 1.1 (*Convergence*) A numerical method is convergent if

$$\||(u - u_N, p - p_N)\|| \to 0 \text{ as } N \to \infty.$$

More precisely,

$$\forall \varepsilon > 0, \exists N_0 > 0, \forall N > N_0 \text{ such that } \||(u - u_N, p - p_N)\|| \le \varepsilon.$$

Chapter 2
FVMs for the Stationary Stokes Equations

Abstract In this chapter, we mainly present the construction and analysis of the stabilized lower order finite volume methods (FVMs) and adaptive finite volume methods for the stationary Stokes equations. An optimal convergence rate for stabilization of lower order finite volume methods is obtained with the same order as that of the corresponding finite element methods. In addition, after proving a number of technical lemmas (such as weighted L^2-norm estimates for regularized Green's functions associated with the Stokes problem), optimal error estimates for the finite volume methods in the L^∞-norm are derived for velocity gradient and pressure without a logarithmic factor $O(|\log h|)$ for the stationary Stokes equations. Finally, convergence of the adaptive stabilized mixed finite volume methods is discussed.

2.1 Introduction

Finite difference (FD), finite element (FE), and finite volume (FV) methods are three major numerical tools for solving partial differential equations. Among these methods, the finite volume method is the most flexible for conservation law because it is based on local conservation over volumes (control volumes or co-volumes). According to the definition of the finite volume method, volume integrals for a partial differential equation that contains a divergence term are converted into surface integrals by using the divergence theorem. These terms are then approximated by numerical fluxes at the surface of each finite volume. Because a flux entering into a volume is identical to that leaving an adjacent volume sharing a common face, this method is conservative. In addition, it can easily be formulated to allow for the use of unstructured meshes to deal with complicated geometries.

This method is a numerical technique that lies somewhere between the finite element and finite difference methods; it has a flexibility similar to that of the finite element method for handling complicated geometries, and its implementation is

© The Author(s), under exclusive license to Springer Nature Switzerland AG 2022
J. Li et al., *Finite Volume Methods for the Incompressible Navier–Stokes Equations*,
SpringerBriefs in Mathematical Methods,
https://doi.org/10.1007/978-3-030-94636-4_2

comparable to that of the finite difference method. Moreover, its numerical solutions usually have certain conservation features that are desirable in many practical applications.

There are a lot of modern finite volume methods available in the literature, including the MAC scheme [50, 51, 53, 56, 61, 64], discrete duality finite volume schemes [40, 41, 62, 65], gradient schemes [45, 52, 66], mixed finite volumes [33, 36, 60, 63], and mimetic finite difference schemes [19, 124, 125]. Most of these schemes are able to deal with very general grids (even non-conforming grids). In this book, we are mainly interested in the finite volume methods, which are also termed the control volume methods, the co-volume methods, or the first-order generalized difference methods. Many papers were devoted to their error analysis for second-order elliptic and parabolic partial differential problems [20, 27, 28, 34, 90]. Error estimates of optimal order in the H^1-norm for the finite volume methods are the same as those for the linear finite element method [27, 49, 83, 98]. Error estimates of optimal order in the L^2-norm can be obtained as well [28, 49]. The finite volume methods for the generalized Stokes problems were studied by many people [31, 32, 35, 114, 130, 134]. They analyzed these methods by using a relationship between them and the finite element methods, and obtained their error estimates through those known for the latter methods.

In this chapter, we study the finite volume methods for the Stokes equations approximated by some attractive finite element pairs through their relationship with the conforming elements. In particular, these pairs are efficient for a saddle-point problem in terms of parallel and multigrid implementation because of the same grid partition. Nevertheless, they do not satisfy the discrete *inf-sup* condition, including the following examples [37, 69, 81, 84, 91, 94, 120]: the (bi)linear velocity-constant pressure ($P_1 - P_0$ or $Q_1 - P_0$) and the (bi)linear velocity-(bi)linear pressure ($Q_1 - Q_1$ or $P_1 - P_1$) pairs that notoriously suffer from the stability problem. The corresponding finite volume results related to the Q_1 can be referred to a few recent papers [100, 101].

We mainly focus on two ways to stabilize these attractive finite element pairs for the stationary Stokes equations.

- Stabilized methods: The main objective is to compensate the negative effect on the discrete *inf-sup* condition (also known as the classical Babuška–Brezzi inequality) and furthermore satisfy the weak coerciveness for these families of mixed lower order finite element pairs that do not satisfy this discrete condition [3, 11, 14, 18, 42, 75–77, 107, 118, 119].
- A technique called a "macroelement condition" [69, 120]: It is applied to enforce this condition. Namely, a P_2 or Q_2 velocity field is approximated on a macro-element mesh obtained by refining $\mathscr{T}_h$ uniformly to obtain the mesh $\mathscr{T}_{h/2}$. These pairs are stable and popular, and this method is called the iso-$P_2 - P_i$, $i = 0, 1$, method.

The key idea in the development and analysis of a stabilized finite volume method for the Stokes equations is to treat it as a generalized Galerkin method or a perturbation of the finite element method. In particular, its relationship with the lower order finite element pairs stabilized by the technique considered above is established. Using this relationship, error estimates of optimal order for the mixed finite volume method are obtained for the Stokes equations [84].

Recently, the adaptive finite volume methods have become very popular among the engineering community for flow computations. There are theoretical frameworks to derive rigorous a posteriori error estimates for finite volume methods; see, e.g., the works of Ern and Vohralík on a posteriori error estimation by equilibrated fluxes [21, 24, 128, 129]. Although the finite volume method is in many cases the method of choice for reliable computations, a theoretical analysis for its adaptive mesh refinement techniques is still not well developed, mainly due to the lack of appropriate a posteriori estimates and complete theoretical results of this method. Compared with the adaptive finite element methods, the test and trial functions of the adaptive finite volume methods belong to different finite dimensional spaces, and a theoretic analysis for the adaptive finite volume methods is far behind that for the adaptive finite element methods. Moreover, the integration cells in the finite volume methods are constructed as dual cells of a primal mesh and, consequently, dual cells of a finer mesh are not subcells of coarser mesh dual cells. There are a few theoretical studies available for the adaptive finite volume methods for the incompressible flow in the literature at present; these relevant studies are the basis and key to establish a series of integrated theories for practical models [48, 128, 129].

This chapter is organized as follows: In the next section, we introduce some notation, the stationary Stokes equations, and the variational form for the stationary Stokes equations. Then, in the third and fourth sections, the stabilized finite volume methods for the Stokes equations are considered, and a relationship between these methods and finite element methods is introduced. Existence, uniqueness, stability, and optimal order priori estimates for the finite volume methods are obtained in the fourth and fifth sections. In the sixth section, a residual type of a posteriori error estimator is designed and studied with the derivation of upper and lower bounds between the exact solution and a finite volume solution. Convergence of the adaptive stabilized mixed finite volume methods is established in the seventh section. Finally, conclusions are drawn in the last section.

2.2 Weak Formulation

We consider the stationary Stokes equations: With $\Omega \subset R^d, d = 2, 3$, being a bounded open set with the Lipschitz boundary $\partial\Omega$,

$$-\mu \Delta u + \nabla p = f \quad \text{in } \Omega, \tag{2.1}$$

$$\text{div } u = 0 \quad \text{in } \Omega, \tag{2.2}$$

$$u = 0 \quad \text{on } \partial\Omega, \tag{2.3}$$

where u represents the velocity vector, p the pressure, f the prescribed body force, and $\mu > 0$ the viscosity.

For the mathematical setting of the Stokes equations, we introduce the following Hilbert spaces:

$$X = [H_0^1(\Omega)]^d, \quad Y = [L^2(\Omega)]^d, \quad M = \left\{ q \in L^2(\Omega) : \int_\Omega q \, dx = 0 \right\},$$

$$V = \{v \in X : \mathrm{div}\, v = 0\}, \quad H = \{v \in Y : \mathrm{div}\, v = 0, \, v \cdot n|_{\partial\Omega} = 0\},$$

$$Z = [L^{3/2}(\Omega)]^d, \quad X' = [H^{-1}(\Omega)]^d, \quad \bar{X} = X \times M, \quad D(A) = [H^2(\Omega)]^d \cap X.$$

Here, the Stokes operator $A : D(A) \to H$ is defined by $A = -P\Delta$ and $P : [L^2(\Omega)]^d \to H$ is the standard L^2-orthogonal projection. The spaces Y is endowed with the L^2-scalar product and the L^2-norm. The space X is equipped with the usual scalar product $(\nabla u, \nabla v)$ and norm $|\cdot|_1$ or $\|\cdot\|_1$, as appropriate.

Especially the standard definitions used for the Sobolev spaces $W_0^{m,r}(\Omega)$ are equivalent, with norm $\|\cdot\|_{m,r}$ and seminorm $|\cdot|_{m,r}$, $m, r > 0$. Also, we define the norm in $\bar{X}$:

$$\||(v, q)\|| = (\|v\|_1^2 + \|q\|_0^2)^{1/2}, \quad (v, q) \in X \times M.$$

The weak formulation associated with (2.1)–(2.3) is to seek $(u, p) \in X \times M$ such that

$$\mathcal{B}((u, p); (v, q)) = (f, v), \qquad \forall (v, q) \in X \times M, \tag{2.4}$$

where the bilinear form

$$\mathcal{B}((u, p); (v, q)) = a(u, v) - d(v, p) + d(u, q), \, \forall (u, p), (v, q) \in X \times M$$

with

$$a(u, v) = (-\Delta u, v) = (\nabla u, \nabla v), \quad d(v, p) = (\mathrm{div}\, v, p).$$

Obviously, the bilinear form $a(\cdot, \cdot)$ is continuous and coercive on the space pair $X \times X$.

We recall the next lemma, which plays an important role in the classical analysis [4].

Lemma 2.1 *For a bounded Lipschitz connected domain Ω and for any $q \in L^2(\Omega)$, there exist a positive constant $C_0 > 0$ and $v \in [H^1(\Omega)]^d$ satisfying*

$$\mathrm{div}\, v = q, \, s.t. \, \|v\|_1 \le C_0 \|q\|_0. \tag{2.5}$$

Furthermore, for any $q \in L^2(\Omega)$ whose integral over Ω is zero, there exists a vector field $v \in [H^1(\Omega)]^d$ that vanishes on the boundary satisfying (2.5).

Proof The reader may refer to Ref. [4], where the authors analyze various situations in detail.   □

Next, the bilinear form $d(\cdot, \cdot)$ is continuous and satisfies the *inf-sup* condition [26, 59, 122]: There exists a positive constant $\beta > 0$ such that, for all $q \in M$,

$$\sup_{0 \neq v \in X} \frac{|d(v, q)|}{\|v\|_1} \geq \beta_0 \|q\|_0, \tag{2.6}$$

where β_0 is a positive constant depending only on Ω. For convenience, we use the same constant β in (2.14) below.

Since the bilinear forms $a(\cdot, \cdot)$ and $d(\cdot, \cdot)$ are continuous, the bilinear form $B(\cdot, \cdot)$ is also continuous. That is,

$$\mathscr{B}((u, p); (v, q)) \leq C(\|u\|_1 + \|p\|_0)(\|v\|_1 + \|q\|_0), \quad \forall (u, p), (v, q) \in X \times M. \tag{2.7}$$

Note that the generic positive constant C (with or without a subscript) depends only on Ω. Subsequently, the constant $\kappa > 0$ (with or without a subscript) will depend only on the data (μ, Ω, f). Moreover, there holds the following inf-sup property:

$$\beta(\|u\|_1 + \|p\|_0) \leq \sup_{(v, q) \in X \times M} \frac{|\mathscr{B}((u, p); (v, q))|}{\|v\|_1 + \|q\|_0}. \tag{2.8}$$

Furthermore, detailed results on existence and uniqueness of a solution to Eqs. (2.1)–(2.3) can be easily obtained ([122], Theorem 2.1).

2.3 Galerkin FV Approximation

Let $\mathscr{T}_h$ be a regular triangular grid in two dimensions or tetrahedral grid in three dimensions [26, 37, 108]. The parameter h is related to a partition of the grid. We set $h_K = diam(K)$, for each $K \in \mathscr{T}_h$, where $diam(K) = \max_{x, y \in K} |x - y|$ is the diameter of element K. Furthermore, $h = \max_{K \in T_h} h_K$. Moreover, we will impose that the grid satisfies the following regularity condition: For a suitable $C > 0$, it holds

$$\frac{h_K}{\rho_K} \leq C, \tag{2.9}$$

where ρ_K is the diameter of the circle inscribed in triangle K.

The partitioned domain $\Omega_h = int \left(\bigcup_{K \in \mathscr{T}_h} K \right)$ represented by the union of the elements in $\mathscr{T}_h$ perfectly coincides with Ω. In this book, we do not consider finite element grids in a non-polygonal domain. From now on, we will not distinguish between Ω and its approximation domain Ω_h, and adopt the symbol Ω to denote the latter below.

Here, we only consider the finite element spaces

$$X_h = \{v \in X : v|_K \in [P_1(K)]^d \ \forall K \in \mathscr{T}_h\},$$

and

$$M_h = \begin{cases} M_h^0 \equiv \{q \in M : q|_K \in P_0(K), \ \forall K \in \mathscr{T}_h\}, \\ M_h^1 \equiv \{q \in M : q|_K \in P_1(K), \ \forall K \in \mathscr{T}_h\}, \end{cases}$$

where P_1 and P_0 denote the spaces of polynomials of degree 1 or 0, respectively. The two-dimensional related quadrilateral case can be found in [11, 100]. With the fundamental theory developed, many experts begin to pay close attention to an attractive computational feature treatment of general elements including non-convex and degenerate elements [5, 6, 10, 12, 39, 44, 54, 55, 99, 111–113].

Let I_h and J_h be two interpolation operators from $X \cap [C^0(\bar{\Omega})]^d$ and M into X_h and M_h, respectively, such that, for $v \in [H^2(\Omega)]^d$ and $q \in H^1(\Omega) \cap M$ [37, 59, 122, 123],

$$\|v - I_h v\|_i \le Ch^{2-i}|v|_2, \quad \|q - J_h q\|_0 \le Ch|q|_1, \quad i = 0, 1. \quad (2.10)$$

In particular, the interpolation operator I_h satisfies

$$\|I_h v\|_1 \le C\|v\|_1, \quad (2.11)$$

where the positive constant $C > 0$ depends on Ω and the grid regularity parameter. Due to the quasi-uniformness of the triangulation $\mathscr{T}_h$, the inverse inequality holds for $v_h \in X_h$:

$$\|v_h\|_1 \le C_1 h^{-1}\|v_h\|_0, \quad (2.12)$$

and

$$\|v_h\|_{L^\infty} \le \begin{cases} C_2 |\log \frac{1}{h}|^{1/2}\|v_h\|_1, & d = 2, \\ C_2 h^{-1/2}\|v_h\|_1, & d = 2, 3. \end{cases} \quad (2.13)$$

In general, we usually apply the estimate with $|\log \frac{1}{h}|$ in two dimensions since $|\log \frac{1}{h}|^{1/2}$ is more accurate than $h^{-1/2}$. Also, the reader can consult the original proof ([123], Lemma 5.1) for more details.

In the present chapter, we formulate the stabilized mixed finite element methods for the lower order pairs $X_h \times M_h^i$, $i = 0, 1$. These two pairs are unstable and do not satisfy the so-called discrete *inf-sup* stability condition:

$$\sup_{0 \ne v_h \in X_h} \frac{d(v_h, q_h)}{\|v_h\|_1} \ge \beta_0 \|q_h\|_0, \quad (2.14)$$

where $\beta_0 > 0$ is independent of h.

In order to highlight the focus of finite volume approximations for the stationary Stokes equations, efficient stabilized methods called the local pressure projection methods have been introduced by adding a symmetrical and compensatory term $S(\cdot, \cdot)$ in a discrete variational formulation: The stabilized bilinear form (cf., [11, 17, 91]) is defined as follows:

$$S(p_h, q_h) = (p_h - \Pi_h p_h, q_h - \Pi_h q_h), \tag{2.15}$$

where Π_h is the elementwise L^2-projection:

$$\Pi_h = \begin{cases} \Pi_0 : L^2(\Omega) \to M_h^1, \\ \Pi_1 : L^2(\Omega) \to M_h^0. \end{cases}$$

Here, the operators Π_0 and Π_1 are applied to stabilizing the lowest equal-order pair $X_h \times M_h^1$ and the lowest order pair $X_h \times M_h^0$, respectively. Furthermore, the mapping satisfies the following properties [11, 91]:

$$\|\Pi_h p\|_0 \le C \|p\|_0, \ \forall p \in M_h, \tag{2.16}$$
$$\|p - \Pi_h p\|_0 \le Ch \|p\|_1, \ \forall p \in H^1(\Omega) \cap M_h. \tag{2.17}$$

Then the bilinear form of the finite element method on $(X_h, M_h) \times (X_h, M_h)$ is given by

$$\mathscr{B}_h((u_h, p_h), (v_h, q_h)) = a(u_h, v_h) - d(v_h, p_h) + d(u_h, q_h) + S(p_h, q_h). \tag{2.18}$$

It can be shown (Theorem 3.1 in [91] and Theorem 4.1 in [11]) that this bilinear form satisfies the continuity and weak coercivity:

$$|\mathscr{B}_h((u_h, p_h), (v_h, q_h))| \le C \left(\|u_h\|_1 + \|p_h\|_0 \right) \left(\|v_h\|_1 + \|q_h\|_0 \right), \tag{2.19}$$
$$\sup_{(v_h, q_h) \in X_h \times M_h} \frac{|\mathscr{B}_h((u_h, p_h), (v_h, q_h))|}{\|v_h\|_1 + \|q_h\|_0} \ge \beta_1 \left(\|u_h\|_1 + \|p_h\|_0 \right), \tag{2.20}$$

where β_1 is independent of h.

The corresponding discrete variational formulation of (2.4) for the Stokes equations is recast: Find $(\bar{u}_h, \bar{p}_h) \in X_h \times M_h$ such that

$$\mathscr{B}_h((\bar{u}_h, \bar{p}_h), (v_h, q_h)) = (f, v_h), \ \forall (v_h, q_h) \in X_h \times M_h. \tag{2.21}$$

Because of (2.19) and (2.20), system (2.21) has a unique solution. Moreover, the error estimate for the finite element solution (u_h, p_h) holds (Theorem 3.1 in [91] and Theorem 4.1 in [11]):

$$\|u - \bar{u}_h\|_0 + h \left(\|u - \bar{u}_h\|_1 + \|p - \bar{p}_h\|_0 \right) \le \kappa h^2 \left(\|u\|_2 + \|p\|_1 + \|f\|_0 \right). \tag{2.22}$$

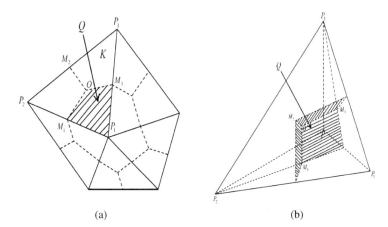

Fig. 2.1 Control volume: **a** 2d case, **b** 3d case

Now, we are in a position to introduce the stabilized finite volume methods. Here, we follow [22, 31, 34, 133, 134] to obtain a dual partition $\mathscr{T}_h^*$ in two- and three-dimensional cases. Let $\mathscr{N}_h$ be the set of all interior vertices of the triangulation $\mathscr{T}_h$ and $\mathscr{E}_h$ the set of all interior edges. Set N_h to be the total number of the nodes. To define the finite volume methods, a dual mesh $\mathscr{T}_h^*$ is introduced based on $\mathscr{T}_h$; the elements V in $\mathscr{T}_h^*$ are called control volumes. The dual mesh can be constructed by the following rule: For each element $K \in \mathscr{T}_h$ with vertices P_j, $j = 1, 2, 3$, select its barycenter O and the midpoint M_j on each of the edges of K, and construct the control volumes in $\mathscr{T}_h^*$ by connecting O to M_j as shown in Fig. 2.1.

Now, the dual finite element space is defined as

$$X_h^* = \left\{ v \in Y : v|_V \in P_0(V), \ v|_{\partial\Omega} = 0, \ \forall V \in \mathscr{T}_h^* \right\}.$$

Obviously, the dimensions of X_h and X_h^* are the same. Furthermore, there exists an invertible linear mapping $\Gamma_h : X_h \to X_h^*$ such that, for

$$v_h(x) = \sum_{j=1}^{N_h} v_h(P_j)\phi_j(x), \qquad x \in \Omega, \ v_h \in X_h,$$

$$v_h^*(x) \equiv \Gamma_h v_h^*(x) = \sum_{j=1}^{N_h} v_h(P_j)\chi_j(x), \qquad x \in \Omega, \ v_h \in X_h,$$

where $\{\phi_j\}$ indicates the basis for the finite element space X_h and $\{\chi_j\}$ denotes the basis for the finite volume space X_h^* that are the characteristic functions associated with the dual partition $\mathscr{T}_h^*$:

$$\chi_j(x) = \begin{cases} 1 & \text{if } x \in V_j \in \mathcal{T}_h^*, \\ 0 & \text{otherwise.} \end{cases}$$

The above idea of connecting the trial and test spaces in the Petrov–Galerkin method through the mapping Γ_h was first introduced in [7, 82, 83] in the context of elliptic problems. Furthermore, the mapping Γ_h satisfies the following properties due to Li [131], and will be used over and over again in this book.

Lemma 2.2 *Let $K \in \mathcal{T}_h$. If $v_h \in X_h$ and $1 \leq r \leq \infty$, then we have*

$$\int_K (v_h - v_h^*)dx = 0. \tag{2.23}$$

Also, they hold:

$$\|v_h - v_h^*\|_{0,r,K} \leq C_3 h_K \|v_h\|_{1,r,K},$$
$$\|v_h - v_h^*\|_{0,r,\partial K} \leq C_3 h_K^{1-1/r} \|v_h\|_{1,r,K}, \tag{2.24}$$
$$\|v_h^*\|_0 \leq C_4 \|v_h\|_0, \tag{2.25}$$

where h_K is the diameter of element K.

Multiplying Eq. (2.1) by $v_h^* \in X_h^*$ and integrating over the dual elements $V \in \mathcal{T}_h^*$, Eq. (2.2) by $q_h \in M_h$ and over the primal elements $K \in \mathcal{T}_h$, and applying Green's formula, we define the following bilinear forms for the finite volume methods:

$$A(u_h, v_h^*) = -\sum_{j=1}^{N_h} v_h(P_j) \cdot \int_{\partial V_j} \frac{\partial u_h}{\partial n} ds, \quad u_h, \, v_h \in X_h,$$

$$D(v_h^*, p_h) = -\sum_{j=1}^{N_h} v_h(P_j) \cdot \int_{\partial V_j} p_h n \, ds = \sum_{j=1}^{N_h} v_h(P_j) \cdot \int_{V_j} \nabla p_h \, ds, \quad p_h \in M_h,$$

$$(f, v_h^*) = \sum_{j=1}^{N_h} v_h(P_j) \cdot \int_{V_j} f \, dx, \quad v_h \in X_h,$$

where n is the unit normal outward to ∂V_j. For the finite volume approximations, we define the bilinear form $\mathscr{C}_h(\cdot, \cdot)$ on $X_h \times M_h \times X_h \times M_h$:

$$\mathscr{C}_h((u_h, p_h), (v_h, q_h)) = A(u_h, v_h^*) + D(v_h^*, p_h) + d(u_h, q_h) + S(p_h, q_h) \tag{2.26}$$

Then, the stabilized finite volume methods are defined for the solution $(u_h, p_h) \in X_h \times M_h$ as follows:

$$\mathscr{C}_h((u_h, p_h), (v_h, q_h)) = (f, v_h^*), \quad \forall (v_h, q_h) \in X_h \times M_h. \tag{2.27}$$

Below we will concentrate on the study of a relationship between the finite element and finite volume methods for the Stokes equations. Here, we extent this relationship for the lowest equal-order finite element pair ([84], Theorem 3.2 and [134], Theorem 2.1) to a wider range of the lower order finite element pairs and we present in this book a different approach inspired from [31, 35, 134].

Lemma 2.3 (Equivalence) *It holds that*

$$A(u_h, v_h^*) = a(u_h, v_h) \quad \forall u_h, \ v_h \in X_h, \tag{2.28}$$

with the following properties:

$$A(u_h, v_h^*) = A(v_h, u_h^*),$$
$$|A(u_h, v_h^*)| \leq \mu \|u_h\|_1 \|v_h\|_1,$$
$$|A(v_h, v_h^*)| \geq \mu \|v_h\|_1^2.$$

Moreover, the bilinear form $D(\cdot, \cdot)$ satisfies

$$D(v_h^*, q_h) = -d(v_h, q_h) \quad \forall (v_h, q_h) \in X_h \times M_h. \tag{2.29}$$

Proof Equation (2.28) was shown in [134] (see Theorem 2.1). For completeness, we will prove (2.28). In fact, noting that $\text{div}(\nabla u_h) = 0$ for any $u_h \in X_h$, using the divergence theorem yields

$$A(u_h, v_h^*) = -\sum_{j=1}^{N_h} v_h(P_j) \cdot \int_{\partial V_j} \frac{\partial u_h}{\partial n} ds, \quad u_h, \ v_h \in X_h,$$

$$= \sum_{j=1}^{N_h} v_h(P_j) \cdot \int_{\partial K_j} \frac{\partial u_h}{\partial n} ds, \quad u_h, \ v_h \in X_h.$$

By the Green formula and (2.23),

$$A(u_h, v_h^*) = \sum_{j=1}^{N_h} (v_h(P_j) - v_h) \cdot \int_{\partial K_j} \frac{\partial u_h}{\partial n} ds + \sum_{K \in \mathscr{T}_h} \int_{\partial K_j} \frac{\partial u_h}{\partial n} v_h ds, \quad u_h, \ v_h \in X_h,$$

$$= \sum_{K \in \mathscr{T}_h} \int_{K_j} \nabla u_h \nabla v_h dx$$

$$= a(u_h, v_h).$$

In the following, it suffices to prove (2.29). For $(v_h, q_h) \in X_h \times M_h$, it follows from the definition of $D(\cdot, \cdot)$ and Green's formula that

$$D(v_h^*, q_h) = \sum_{j=1}^{N} \int_{V_j} \nabla q_h \cdot v_h^* \, dx$$

$$= \sum_{j=1}^{N} \sum_{K \in \mathcal{T}_h} \int_{K \cap V_j} \nabla q_h \cdot v_h^* \, dx$$

$$= \sum_{K \in \mathcal{T}_h} \nabla q_h \cdot \int_{K} v_h \, dx$$

$$= -(\operatorname{div} v_h, q_h) = -d(v_h, q_h),$$

in the case of $P_1 - P_1$ and

$$D(v_h^*, q_h) = \sum_{j=1}^{N} \int_{V_j} v_h^* \cdot n p_h ds$$

$$= -\sum_{j=1}^{N} \sum_{K \cap V} \int_{\partial K} v_h^* \cdot n p_h ds$$

$$= -\sum_{K \in \mathcal{T}_h} p_h \int_{\partial K} v_h^* \cdot n ds$$

$$= -\int_{\partial K} v_h \cdot n p_h ds$$

$$= -d(v_h, p_h).$$

in the case of $P_1 - P_0$. These results imply (2.29).   □

2.4   Existence and Uniqueness Theorems

Applying Lemmas 2.2 and 2.3, the next result establishes the continuity and weak coercivity of the finite volume approximations for the stationary Stokes equations.

Theorem 2.1 *It holds that*

$$|\mathscr{C}_h((u_h, p_h), (v_h, q_h))| \le C \left(\|u_h\|_1 + \|p_h\|_0\right) \left(\|v_h\|_1 + \|q_h\|_0\right)$$
$$\forall (u_h, p_h), \ (v_h, q_h) \in X_h \times M_h. \quad (2.30)$$

Moreover,

$$\sup_{(v_h, q_h) \in X_h \times M_h} \frac{|\mathscr{C}_h((u_h, p_h), (v_h, q_h))|}{\|v_h\|_1 + \|q_h\|_0} \ge \beta^* \left(\|u_h\|_1 + \|p_h\|_0\right)$$
$$\forall (u_h, p_h) \in X_h \times M_h, \quad (2.31)$$

where the positive constant β^* is independent of h.

Proof It follows from the equivalence Lemma 2.3 that

$$|\mathscr{C}_h((u_h, p_h), (v_h, q_h))| \le |A(u_h, v_h^*)| + |D(v_h^*, p_h)| + |d(u_h, q_h)| + |S(p_h, q_h)|$$
$$\le C \left(\|u_h\|_1 + \|p_h\|_0\right) \left(\|v_h\|_1 + \|q_h\|_0\right);$$

i.e., the continuity result (2.30) holds.

By Lemma 3.1, for each $p_h \in M_h \subset M$, there exists $w \in X$ [14, 26] such that

$$\mathrm{div}\, w = p_h$$

and

$$\|w\|_1 \le C_0 \|p_h\|_0, \qquad d(w, p_h) \le C \|w\|_1 \|p_h\|_0. \quad (2.32)$$

Let $w_h = I_h w \in X_h$, which satisfies (2.10) and (2.11). Then it follows from Young's inequality that

$$a(u_h, w_h) \le \mu \|u_h\|_1 \|w_h\|_1$$
$$\le C_0 \mu \|u_h\|_1 \|p_h\|_0$$
$$\le \frac{1}{4} \|p_h\|_0^2 + C_0^2 \mu^2 \|u_h\|_1^2. \quad (2.33)$$

The proof of the term $d(w_h, p_h)$ is separated into two cases. First, setting $\kappa_1 = CC_0C_1$, letting the finite element pair be $P_1 - P_1$, noting that ∇p_h is constant on each K, and using the inverse inequality and (2.32), we obtain

$$
\begin{aligned}
d(w_h, p_h) &= d(w, p_h) - d(w - w_h, p_h) \\
&\geq \|p_h\|_0^2 - Ch\|w\|_1\|\nabla(p_h - \Pi_1 p_h)\|_0 \\
&\geq \|p_h\|_0^2 - CC_1\|w\|_1\|p_h - \Pi_1 p_h\|_0 \\
&\geq \|p_h\|_0^2 - CC_0C_1\|p_h\|_0\|p_h - \Pi_1 p_h\|_0 \\
&\geq \frac{3}{4}\|p_h\|_0^2 - \kappa_1^2\|p_h - \Pi_1 p_h\|_0^2.
\end{aligned}
\tag{2.34}
$$

On the other hand, letting the finite element pair be $P_1 - P_0$, using $\nabla p_h = \mathbf{0}$, $[\Pi_1 p_h] = 0$, Green's formula, the Young inequality, the inverse inequality

$$
\|\nabla q_h\|_0 + h^{-1/2} \sum_{K \in \mathcal{T}_h} \|[q_h]\|_{0,\partial K} \leq C_1 h^{-1}\|q_h\|_0 \quad \forall q_h \in Q_h,
$$

and noting that each interior face ∂K participates twice in the sum, we see that

$$
\begin{aligned}
d(w - w_h, p_h) &= \sum_K \int_\Omega \operatorname{div}(w - w_h) p_h dx \\
&= \sum_K \int_{\partial K} [p_h](w - w_h) \cdot n ds \\
&= \sum_K \left(\int_{\partial K} |[p_h]|^2 ds \right)^{1/2} \left(\int_{\partial K} |w - w_h|^2 ds \right)^{1/2} \\
&\leq Ch^{1/2} \sum_{K \in \mathcal{T}_h} \|[p_h]\|_{0,2,\partial K}\|w\|_1 \\
&\leq CC_0 h^{1/2} \sum_K \|[p_h - \Pi_0 p_h]\|_{0,2,\partial K}\|p_h\|_0
\end{aligned}
\tag{2.35}
$$

yields

$$
\begin{aligned}
d(w_h, p_h) &= d(w, p_h) - d(w - w_h, p_h) \\
&\geq \|p_h\|_0^2 - CC_0 h^{1/2} \sum_K \|[p_h - \Pi_0 p_h]\|_{0,2,\partial K}\|p_h\|_0 \\
&\geq \|p_h\|_0^2 - CC_0C_1\|p_h - \Pi_0 p_h\|_0\|p_h\|_0 \\
&\geq \frac{3}{4}\|p_h\|_0^2 - \kappa_1^2\|p_h - \Pi_0 p_h\|_0^2,
\end{aligned}
\tag{2.36}
$$

where C_1 is defined in (2.12).

Now, choosing $(v_h, q_h) = (u_h - \alpha w_h, p_h)$ with a positive constant α (to be determined below) in the bilinear form $\mathscr{C}_h(\cdot, \cdot)$ and applying the equivalence Lemma 2.3 and Young's inequality, we see that

$$\mathscr{C}_h((u_h, p_h), (u_h - \alpha w_h, p_h))$$
$$= A(u_h, u_h^*) - \alpha A(u_h, w_h^*) - \alpha D(w_h^*, p_h) + S(p_h, p_h)$$
$$\geq \mu(1 - \alpha C_0^2 \mu) \|u_h\|_1^2 + \frac{\alpha}{2} \|p_h\|_0^2 + \left(1 - \alpha \kappa_1^2\right) \|p_h - \Pi_h p_h\|_0^2. \quad (2.37)$$

Obviously, choosing $\alpha = \min \left\{ \dfrac{1}{2C_0^2\mu}, \dfrac{1}{2\kappa_1^2} \right\}$ gives

$$\min \left\{1 - \alpha C_0^2 \mu, 1 - \alpha \kappa_1^2 \right\} \geq \frac{1}{2}.$$

With this choice, (2.37) leads to

$$|\mathscr{C}_h((u_h, p_h), (u_h - \alpha w_h, p_h))| \geq \kappa_2 \left(\|u_h\|_1 + \|p_h\|_0\right)^2. \quad (2.38)$$

Also, it is clear, with (2.32) and the choice of α, that

$$\|\nabla(u_h - \alpha w_h)\|_0 + \|p_h\|_0 \leq \|u_h\|_1 + \alpha\|w_h\|_1 + \|p_h\|_0$$
$$\leq \kappa_3 \left(\|u_h\|_1 + \|p_h\|_0\right). \quad (2.39)$$

Finally, by setting $\beta_2 = \kappa_2/\kappa_3$ and combing (2.38) with (2.39), we complete the proof of (2.31).   □

It follows from Theorem 2.1 that the new stabilized finite volume system (2.27) has a unique solution $(u_h, p_h) \in X_h \times M_h$.

2.5 Priori Estimate

In this section, we now obtain optimal L^2, H^1 and L^∞ estimates for the stabilized finite volume methods for the stationary Stokes equations.

2.5.1 Superclose

Here, we analyze a relationship between the solutions of the stabilized finite volume methods and the corresponding finite element methods for the stationary Stokes equations.

Theorem 2.2 (Superclose) *Let* $(\bar{u}_h, \bar{p}_h)$ *and* (u_h, p_h) *be the solutions of* (2.21) *and* (2.27), *respectively. Then*

$$\|\bar{u}_h - u_h\|_1 + \|\bar{p}_h - p_h\|_0 \leq \kappa h^2 \|f\|_1. \tag{2.40}$$

Proof Subtracting (2.21) from (2.27) gives

$$\mathscr{C}_h((\bar{u}_h - u_h, \bar{p}_h - p_h), (v_h, q_h)) = (f, v_h - v_h^*) \quad \forall (v_h, q_h) \in X_h \times M_h. \tag{2.41}$$

Obviously, setting $(e, \eta) = (\bar{u}_h - u_h, \bar{p}_h - p_h)$, we deduce from (2.31) and the equivalence Lemma 2.3 that

$$\sup_{(v_h, q_h) \in X_h \times M_h} \frac{|\mathscr{C}_h((e, \eta), (v_h, q_h))|}{\|v_h\|_1 + \|q_h\|_0} \geq \beta^* (\|e\|_1 + \|\eta\|_0). \tag{2.42}$$

In addition, setting $\hat{\pi}_h f = \frac{1}{|K|} \int_K f \, dx$,

$$\begin{aligned} |(f, v_h - v_h^*)| &= |(f - \hat{\pi}_h f, v_h - v_h^*)| \\ &\leq C h^{1+i} \|f\|_i \|v_h\|_1, \quad i = 0, 1. \end{aligned} \tag{2.43}$$

Combining (2.41)–(2.43), we obtain the desired result for $i = 1$. $\qquad\square$

2.5.2 Optimal Analysis

Note from our discussion on Theorem 2.2 that if $f \in [L^2(\Omega)]^d$, then the relationship between the solutions of the stabilized finite volume methods and the corresponding finite element methods is not super-convergent but optimal. Furthermore, optimal estimates can be derived from the standard Galerkin method and special interpolation $(\bar{u}_h, \bar{p}_h) \in X_h \times M_h$.

Theorem 2.3 (Optimal estimate) *Let* (u, p) *and* (u_h, p_h) *be the solutions of* (2.4) *and* (2.27), *respectively. Then*

$$\|u - u_h\|_1 + \|p - p_h\|_0 \leq \kappa h (\|u\|_2 + \|p\|_1 + \|f\|_0). \tag{2.44}$$

Proof Combining (2.41)–(2.43), we obtain for $i = 0$

$$\|e\|_1 + \|\eta\|_0 \leq C h \|f\|_0, \tag{2.45}$$

which, together with the corresponding finite element results related to the finite
element methods (2.22), gives the desired result (2.44).   □

2.5.3 L^2 Estimate for Velocity

For the proof of our error estimate in the L^2-norm for velocity, we introduce the
following Aubin–Nitsch technique. To obtain an estimate for the error $\chi = u - u_h$
in the L^2-norm using a duality argument, set

$$B((v, q), (\Phi, \Psi)) = a(v, \Phi) + d(v, \Psi) - d(\Phi, q), (v, q), (\Phi, \Psi) \in X \times M.$$
(2.46)

Then we consider the dual problem by seeking $(\Phi, \Psi) \in X \times M$ such that

$$B((v, q), (\Phi, \Psi)) = (v, \chi) \quad \forall (v, q) \in X \times M.$$
(2.47)

When Ω is convex, the solution of (2.47) satisfies the regularity

$$\|\Phi\|_2 + \|\Psi\|_1 \leq \kappa \|\chi\|_0.$$
(2.48)

Theorem 2.4 *Let (u, p) and (u_h, p_h) be the solutions of (2.4) and (2.27),
respectively, and $f \in [H^1(\Omega)]^d$. Then*

$$\|u - u_h\|_0 \leq \kappa h^2 (\|u\|_2 + \|p\|_0 + \|f\|_1).$$
(2.49)

Proof Choosing $(v, q) = (\chi, \varpi) = (u - u_h, p - p_h)$ in (2.47) and $(v, q) = (I_h\Phi, J_h\Psi)$ in (2.4) and (2.27), we see that

$$\|u - u_h\|_0^2 = \mathscr{C}_h((\chi, \varpi), (\Phi - I_h\Phi, \Psi - J_h\Psi)) + (f, I_h\Phi - I_h\Phi^*)$$
$$+ S(p, J_h\Psi).$$
(2.50)

Clearly, it follows from (2.15), (2.16) and Theorem 2.3 that

$$|S(p, J_h\Psi)| \leq Ch^2\|p\|_1\|\Psi\|_1 \leq Ch^2\|p\|_1\|u - u_h\|_0,$$
(2.51)

and

$$|\mathscr{C}_h((\chi, \varpi), (\Phi - I_h\Phi, \Psi - J_h\Psi))|$$
$$\leq (\|\chi\|_1 + \|\varpi\|_0)(\|\Phi - I_h\Phi\|_1 + \|\Psi - J_h\Psi\|_0)$$
$$\leq Ch^2(\|u\|_2 + \|p\|_0 + \|f\|_0)(\|\Phi\|_2 + \|\Psi\|_1) \tag{2.52}$$
$$\leq Ch^2(\|u\|_2 + \|p\|_0 + \|f\|_0)\|u - u_h\|_0.$$

In addition, setting $\Phi^* = \Gamma_h(I_h\Phi)$ and using (2.43) and (2.24) yield

$$|(f, I_h\Phi - I_h\Phi^*| = |(f - \hat{\pi}_h f, I_h\Phi - I_h\Phi^*)|$$
$$\leq \|f - \hat{\pi}_h f\|_0 \|I_h\Phi - I_h\Phi^*\|_0$$
$$\leq Ch^2\|f\|_1\|I_h\Phi\|_1$$
$$\leq Ch^2\|f\|_1\|u - u_h\|_0, \tag{2.53}$$

which, together with (2.50)–(2.52), yields the desired result (2.49). □

Remark In fact, Theorem 2.4 can be directly derived from the superclose results and (2.22).

2.5.4 Optimal L^∞ Estimate

In this subsection, a novel L^∞ analysis for finite volume approximations of the Stokes problem is presented. Optimal estimates in the L^∞-norm for velocity gradient and pressure of this problem are obtained by using new technical results. In particular, these optimal estimates for the finite volume approximations are obtained without the presence of a usual logarithmic factor $O(|\log h|)$ for the stationary Stokes problem.

Here, the aim is to give a stability and optimal convergence analysis in the L^∞-norm for velocity gradient and pressure, which is not available in the literature for the finite volume approximations of the stationary Stokes equations. The main difficulty of the convergence analysis is to obtain optimal error estimates in this norm by removing the logarithmic factor $O(|\log h|)$ that appears in the traditional estimates. The analysis in this subsection is based on a technique using the weighted Sobolev norms introduced in [13, 46, 57] for the finite element approximations of the Stokes equations. Duran et al. [46] provided a sharp L^∞-norm error estimate for the finite element approximations of the Stokes problem with the logarithmic factor. Girault et al. [57, 58] adapted the analysis of the finite element methods in [13, 46] to remove the logarithmic factor by working with the weight $\sigma^{\mu/2}$ to be defined below.

In the coming analysis, a technique called a "macroelement condition" [120] is applied to enforce the inf-sup condition for the lower order finite element pairs. These pairs are stable and popular, and this method is also called the iso-$P_2 - P_i$, $i = 0, 1$, method.

The interpolation operators I_h and J_h satisfy the following additional assumption [57]:

Assumption (A3):

- I_h is quasi-local: For all $K \in \mathscr{T}_h$,

$$\|I_h v - v\|_{0,K} + h_K \|\nabla(I_h v - v)\|_{0,K} \le C h_K^2 |v|_{2,\Delta K},$$
$$\|\nabla I_h v\|_{0,K} \le C |v|_{1,\Delta K};$$

- I_h satisfies the discrete divergence-preserved property:

$$d(I_h v - v, q_h) = 0, \qquad \forall q_h \in \overline{M}_h;$$

- J_h is also quasi-local: For all $K \in \mathscr{T}_h$,

$$\|J_h q - q\|_{0,K} \le C h_K |q|_{1,\Delta K}.$$

Here, ΔK is a macro-element containing at most L elements of $\mathscr{T}_h$ including K, with L being a fixed integer independent of h, and the functions in $\overline{M}_h$ are those in M_h without the zero mean-value constraint over Ω. The additional property of quasi-locality is fundamental here for deriving weighted estimates.

We now recall some basic regularity results and properties of Green's function for the Stokes equation. Toward that end, we fix an element of the matrix ∇u_h, e.g., $\frac{\partial u_{h,i}}{\partial x_j}$, and an appropriate point x_0 located in the element $K \in \mathscr{T}_h$, where $\left|\frac{\partial u_{h,i}}{\partial x_j}\right|$ is maximum. An approximate mollifier δ_M supported by K is defined so that

$$D\delta_M = \frac{\partial(\delta_M e_i)}{\partial x_j}, \quad \int_\Omega \delta_M dx = 1, \quad \left\|\frac{\partial u_{h,i}}{\partial x_j}\right\|_{L^\infty} = \left(\delta_M, \frac{\partial u_{h,i}}{\partial x_j}\right), \quad (2.54)$$

where e_i is the unit vector in the i-direction ($i = 1, 2, \ldots, d$). Then we define the regularized Green's functions as follows [13, 110]: Find $(G, Q) \in X \times M$ such that

$$a(G, v) - d(v, Q) + d(G, q) = -(D\delta_M, v) \ \forall \ (v, q) \in X \times M, \quad (2.55)$$

which satisfies the following estimates ([57], Corollary 3.4 and Theorem 3.6):

$$\|\sigma^{\mu/2-1}\nabla G\|_0 + \|\sigma^{\mu/2-1}Q\|_0 \le C h^{\theta/2-1},$$
$$\|\sigma^{\mu/2}\Delta G\|_0 + \|\sigma^{\mu/2}\nabla Q\|_0 \le C h^{\theta/2-1}, \quad (2.56)$$

where $\sigma(x) = \sqrt{|x - x_0|^2 + (\theta_1 h)^2}$ ($|x - x_0| < R$, $R > 0$), $\mu = 2 + \theta$, with $0 < \theta < 1$, and $C > 0$ is independent of the positive constant $\theta_1 > 1$ and the mesh size h.

Correspondingly, a Stokes projection is defined by $(G_h, Q_h) \in X_h \times M_h$ of (G, Q):

$$a(G - G_h, v_h) - d(v_h, Q - Q_h) + d(G - G_h, q_h) = 0$$
$$\forall (v_h, q_h) \in X_h \times M_h, \qquad (2.57)$$

and satisfies the following properties ([57], Theorem 3.11):

$$\|\nabla G_h\|_0 + \|Q_h\|_0 \leq C(\|\nabla G\|_0 + \|Q\|_0),$$
$$\|\sigma^{\mu/2} \nabla (G - G_h)\|_0 + \|\sigma^{\mu/2}(Q - Q_h)\|_0 \leq C h^{\theta/2}. \qquad (2.58)$$

Based on these preparations, we obtain the following optimal L^∞ estimate for the velocity gradient of the Stokes equations.

Lemma 2.4 *Let* (u, p) *and* (u_h, p_h) *be the solutions of* (2.4) *and* (2.27), *respectively. Then it holds that*

$$\|\nabla(u - u_h)\|_{L^\infty} \leq \kappa h(|u|_{2,\infty} + |p|_{1,\infty} + \|f\|_1). \qquad (2.59)$$

Proof Using the definition of the L^∞-norm and taking $(v, q) = (e_h, 0) = (I_h u - u_h, 0)$ in (2.55) yield that

$$\|\nabla e_h\|_{L^\infty} = a(G, e_h) - d(e_h, Q). \qquad (2.60)$$

Due to (2.57) with $q_h = 0$, we obtain

$$\|\nabla e_h\|_{L^\infty} = a(G_h, e_h) - d(e_h, Q_h). \qquad (2.61)$$

Using the macroelement technique, the stabilization term is avoided. Thus, subtracting (2.27) from (2.4), we have the following identity:

$$a(u - u_h, v_h) - d(v_h, p - p_h) + d(u - u_h, q_h) = (f, v_h - v_h^*). \qquad (2.62)$$

Then, setting $E = I_h u - u$ and taking $(v_h, q_h) = (G_h, 0)$ in (2.62) give

$$a(e_h, G_h)$$
$$= a(E, G_h) + d(G_h, p - p_h) + (f, G_h - G_h^*)$$
$$= a(E, G_h - G) + d(G_h - G, p - J_h p_h) + (f, G_h - G_h^*) + a(E, G), \quad (2.63)$$

since

$$d(G_h, p - p_h) = d(G_h - G, p - J_h p).$$

For the last term of the third line in (2.63), using (2.55), we get

$$a(E, G) = a(G, E) = d(E, Q) - (D\delta_M, E). \tag{2.64}$$

Thanks to the second property of **Assumption** $(A3)$ for the interpretation $I_h u$, we see that

$$
\begin{aligned}
d(E, Q) - d(e_h, Q_h) &= d(E, Q) - d(E, Q_h) \\
&= d(E, Q - Q_h).
\end{aligned} \tag{2.65}
$$

Then, substituting (2.63) and (2.65) into (2.61) yields that

$$
\begin{aligned}
\|\nabla e_h\|_{L^\infty} &= a(E, G_h - G) + d(G_h - G, p - J_h p_h) + (f, G_h - G_h^*) \\
&\quad - d(e_h, Q_h) + a(E, G) \\
&= a(E, G_h - G) + d(G_h - G, p - J_h p_h) + (f, G_h - G_h^*) \\
&\quad - d(e_h, Q_h) + d(E, Q) - (D\delta_M, E) \\
&= a(E, G_h - G) + d(G_h - G, p - J_h p_h) - d(E, Q - Q_h) \\
&\quad + (f, G_h - G_h^*) + \|\nabla E\|_{L^\infty},
\end{aligned} \tag{2.66}
$$

For $\|\nabla(G_h - G)\|_{L^1}$, using the Cauchy–Schwartz inequality, we see that

$$
\begin{aligned}
\|\nabla(G_h - G)\|_{L^1} &= \int_\Omega |\nabla(G_h - G)| dx \\
&\leq \left(\int_\Omega \sigma^\mu |\nabla(G_h - G)|^2 dx \right)^{1/2} \left(\int_\Omega \sigma^{-\mu} dx \right)^{1/2},
\end{aligned} \tag{2.67}
$$

where it follows from ([57], Sect. 2) that

$$\int_\Omega \sigma^{-\mu} dx \leq Ch^{-\theta}, \ 0 < \theta < 1, \ \mu = 2 + \theta. \tag{2.68}$$

Thus, by (2.58) and (2.67), we have

$$\|\nabla(G_h - G)\|_{L^1} \leq C. \tag{2.69}$$

The same approach can be used to obtain a bound for the term $\|Q_h - Q\|_{L^1}$. Therefore, using the Cauchy–Schwartz inequality again and following the proof in the L^1-norm in [57], we have

$$|a(E, G_h - G) + d(G_h - G, p - J_h p_h) - d(E, Q - Q_h)|$$
$$\leq C(\|\nabla E\|_{L^\infty} + \|p - J_h p\|_{L^\infty})(\|\nabla(G - G_h)\|_{L^1} + \|Q - Q_h\|_{L^1})$$
$$\leq C(\|\nabla E\|_{L^\infty} + \|p - J_h p\|_{L^\infty})$$
$$\leq Ch(|u|_{2,\infty} + |p|_{1,\infty}). \tag{2.70}$$

As for the term $\|\nabla G\|_0$, using again the Hölder inequality and the first inequality of (2.56), we see that

$$\|\nabla G\|_0^2 \leq \max \sigma^{2-\mu} \int_\Omega \sigma^{\mu-2} |\nabla G|^2 dx$$
$$\leq Ch^{2-\mu} \|\sigma^{\mu/2-1} \nabla G\|_{L^2(\Omega)}^2 \leq Ch^{-2}. \tag{2.71}$$

Similarly,

$$\|Q\|_0 \leq Ch^{-1}. \tag{2.72}$$

Applying (2.23), (2.24), (2.58), (2.71) and the same approach as for (2.43), we see that

$$|(f, G_h - G_h^*)| = |(f - \hat{\pi}_h f, G_h - G_h^*)| \leq Ch\|f\|_1. \tag{2.73}$$

Finally, combing all these inequalities, the triangle inequality, (2.10) and (2.66) yields the desired result. $\square$

As for the analysis of $\|p - p_h\|_{L^\infty}$, a different regularized Green's function is defined as follows [13, 110]: Find $(U, V) \in X \times M$ such that

$$a(U, v) + d(v, V) + d(U, q) = (\delta_M - B, q), \ (v, q) \in X \times M. \tag{2.74}$$

Analogically, the solution of problem (2.74) satisfies ([57], Corollary 3.4)

$$\|\sigma^{\mu/2-1} \nabla U\|_0 + \|\sigma^{\mu/2-1} V\|_0 \leq Ch^{\theta/2-1}. \tag{2.75}$$

Furthermore, we define its Stokes projection $(U_h, V_h) \in X_h \times M_h$ as follows:

$$a(U - U_h, v_h) + d(v_h, V - V_h) - d(U - U_h, q_h) = 0 \ \forall (v_h, q_h) \in X_h \times M_h, \tag{2.76}$$

which satisfies the following estimates ([57], Theorem 3.11):

$$\|\nabla U_h\|_0 + \|V_h\|_0 \leq C(\|\nabla U\|_0 + \|V\|_0),$$
$$\|\sigma^{\mu/2} \nabla (U - U_h)\|_0 + \|\sigma^{\mu/2}(V - V_h)\|_0 \leq Ch^{\theta/2}. \tag{2.77}$$

Let B be a fixed function in $C_0^\infty(\Omega)$ such that $\int_\Omega B(x) \, dx = 1$. Then $\delta_M - B \in L_0^2(\Omega)$. We now prove the following lemma:

Lemma 2.5 *Let (u, p) and (u_h, p_h) be the solutions of (2.4) and (2.27), respectively. Then it holds that*

$$\|p - p_h\|_{L^\infty} \leq Ch(|u|_{2,\infty} + |p|_{1,\infty} + \|f\|_1). \tag{2.78}$$

Proof By setting $(v, q) = (u - u_h, p - p_h)$ in (2.74), we see that

$$(\delta_M - B, p - p_h) = a(U, u - u_h) + d(u - u_h, V) - d(U, p - p_h). \tag{2.79}$$

Also, taking $(v_h, q_h) = (U_h, V_h)$ in (2.62), we obtain

$$a(u - u_h, U_h) - d(U_h, p - p_h) + d(u - u_h, V_h) = (f, U_h - U_h^*). \tag{2.80}$$

Using (2.79) and (2.80), we find that

$$(\delta_M, p - p_h) = a(u - u_h, U - U_h) + d(u - u_h, V - V_h) - d(U - U_h, p - p_h)$$
$$+(B, p - p_h) + (f, U_h - U_h^*). \tag{2.81}$$

Noting that $d(U - U_h, q_h) = 0$ in (2.76) and setting $\eta_h = J_h p - p_h$, the above equality (2.81) can be written as follows:

$$\|\eta_h\|_{L^\infty} = a(u - u_h, U - U_h) + d(u - u_h, V - V_h) - d(U - U_h, p - J_h p)$$
$$+(B, p - p_h) + (\delta_M, J_h p - p)$$
$$+(f, U_h - U_h^*). \tag{2.82}$$

Using the Hölder inequality and (2.10), and the inequality

$$\|\mathrm{div}\, v\|_0 \leq \sqrt{d}\|\nabla v\|_0, \tag{2.83}$$

we see that

$$a(u - u_h, U - U_h) + d(u - u_h, V - V_h) - d(U - U_h, p - J_h p)$$
$$\leq ((1 + \sqrt{d})\|\nabla(u - u_h)\|_{L^\infty} + \|p - J_h p\|_{L^\infty})(\sqrt{d}\|\nabla(U_h - U)\|_{L^1} + \|V_h - V\|_{L^1})$$
$$\leq Ch(|u|_{2,\infty} + |p|_{1,\infty} + \|f\|_1)(\|\nabla(U_h - U)\|_{L^1} + \|V_h - V\|_{L^1}).$$

Thanks to the interpolation property (2.10), an optimal estimate for the corresponding finite element methods for the Stokes equations (2.22), and a simple calculation, we obtain

$$(B, p - p_h) + (\delta_M, J_h p - p)$$
$$\leq C(\|p - p_h\|_0 + \|p - J_h p\|_{L^\infty})$$
$$\leq Ch(\|u\|_2 + |p|_{1,\infty} + \|f\|_0).$$

Furthermore, we apply the same procedure as for (2.67) with respect to $\|\nabla(U - U_h)\|_{L^1}$, $\|V - V_h\|_{L^1}$, $\|\nabla U\|_0$ and $\|V\|_0$ to obtain

$$\|\nabla U\|_0 + \|V\|_0 \leq Ch^{-1}, \quad \|\nabla(U - U_h)\|_{L^1(\Omega)} + \|V - V_h\|_{L^1(\Omega)} \leq C. \ (2.84)$$

Combining all these inequalities, the interpolation property (2.10), (2.73), and a triangle inequality, we can obtain the desired estimate.   □

Remarks (1) In Lemmas 2.4 and 2.5, optimal analysis in the L^∞-norm for velocity gradient and pressure does not require the inf-sup condition or weak coerciveness. (2) Compared with the L^∞ analysis for the Navier–Stokes equations in [87, 88], the analysis here does not require to estimate bounds for $\|\nabla u_h\|_{L^\infty}$ and $\|p_h\|_{L^\infty}$. (3) We note that $\|f\|_1$ is required in the optimal analysis in the L^∞-norm. The main difficulty is that Green's function is bounded by the factor $O(h^{-1})$ and thus a trick is needed to compensate for the negative effect. Moreover, another difficulty arises from losing the symmetry in the Petrov–Galerkin system for the trial and test functions when using a duality argument.

Finally, by Lemmas 2.4 and 2.5, we obtain the following main result:

Theorem 2.5 *Let (u, p) and (u_h, p_h) be the solutions of (2.4) and (2.27), respectively. Then it holds that*

$$\|\nabla(u - u_h)\|_{L^\infty} + \|p - p_h\|_{L^\infty} \leq \kappa h(|u|_{2,\infty} + |p|_{1,\infty} + \|f\|_1). \quad (2.85)$$

Remark In this subsection, optimal L^∞ estimates for finite volume approximations of the Stokes equations are obtained under the assumption of a Lipschitz polygon or polyhedron, satisfying suitable sufficient conditions on the inner angles of its boundary. Recently, Girault et al. [58] modified the weighted L^2 estimates for regularized Green functions, which impose some restrictions on a non-convex domain. Therefore, the reader can easily follow this approach and prove error estimates in the maximum norm in $[W^{1,\infty}(\Omega)]^3 \times L^\infty(\Omega)$ for the incompressible Stokes equations in convex, three-dimensional domains with simplicial boundaries.

2.6 Posteriori Estimation

In this subsection, we will analyze a residual based error estimator for the stabilized finite volume methods for the incompressible flow approximated by the lower order finite element pairs. The upper and lower bounds between the exact solution and the finite volume solution are obtained by using a relationship between the finite element and finite volume methods, and some techniques involving bubble functions are applied in this section as well. For simplicity, we set $\mu = 1$ in this section.

2.6.1 Upper Bound

Here, a Residual-based error estimator is investigated by using the techniques of residual a posteriori error estimates and the stabilized finite volume methods for the Stokes equations.

Let H and h be the mesh scales of two conforming triangulations $\mathcal{T}_H$ and $\mathcal{T}_h$. The error $\||(u - u_H, p - p_H)\||_\Omega$ between (u, p) and (u_H, p_H) is bounded by the global error estimator η_H defined below.

Note that the stabilized finite volume approximation (2.27) is approximated by the lower order elements. Thus, we set

$$R_K = f + \Delta u_H - \nabla p_H$$

on K and for an edge E of a triangle K,

$$J_E = \left[\frac{\partial u_H}{\partial n} - p_H I \cdot n \right]_E,$$

where n is the unit normal outward to $E \in \mathcal{E}_H$. Furthermore, we define the following local error estimator: For any $(v_H, q_H) \in X_H \times M_H$,

$$\eta_H^2(K) = H_K^2 \|R_K\|_{0,K}^2 + \|\operatorname{div} u_H\|_{0,K}^2 + H_E \|J_E\|_{0,E}^2.$$

Then the global error estimator is given by

$$\eta_H = \left(\sum_{K \in \mathcal{T}_H} \eta_H^2(K) \right)^{1/2}.$$

Recalling [84], an inequality is obtained for the Stokes equations approximated by the lowest equal-order finite element. In the following, we extend this inequality for the lower order finite elements. It is an important result for the upper bound of the adaptive stabilized finite volume methods.

Lemma 2.6 *For any $(v, q) \in X \times M$ and $v_h^* \in X_H^*$, it holds that*

$$\mathscr{B}((u - u_H, p - p_H); (v, q))$$
$$= \sum_{K \in \mathscr{T}_H} \int_K (R_K \cdot (v - v_h^*) + \mathrm{div}\, u_H q)dx$$
$$+ \sum_{E \in \mathscr{E}_H} \int_E J_E(v - v_h^*)ds. \tag{2.86}$$

Proof Using the definition of $\mathscr{B}((\cdot, \cdot), (\cdot, \cdot))$ and (2.4) yields

$$\mathscr{B}((u - u_H, p - p_H), (v, q))$$
$$= [a(u, v) - d(v, p) + d(u, q)] - [a(u_H, v) - d(v, p_H) + d(u_H, q)]$$
$$= - \left[\sum_{Q \subset \mathscr{Q}_H} \int_Q (\nabla u_H \cdot \nabla(v - v_H^*) - \mathrm{div}(v - v_H^*)p_H)dx - d(u - u_H, q) \right]$$
$$+ (f, v - v_h^*) + (f, v_h^*). \tag{2.87}$$

By applying the divergence theorem on each triangle, it follows that

$$\sum_{Q \subset \mathscr{Q}_H} \int_Q (\nabla u_H \cdot \nabla(v - v_H^*)dx$$
$$= \sum_{Q \subset \mathscr{Q}_H} \left(\int_{\partial Q} \frac{\partial u_H}{\partial n}(v - v_h^*)ds + \int_Q -\Delta u_H \cdot (v - v_H^*)dx \right). \tag{2.88}$$

Also, we see that

$$\sum_{Q \subset \mathscr{Q}_H} \int_Q \mathrm{div}(v - v_H^*)p_H dx$$
$$= \sum_{Q \subset \mathscr{Q}_H} \left(\int_{\partial Q} p_H(v - v_h^*) \cdot n ds - \int_Q \nabla p_H \cdot (v - v_h^*)dx \right). \tag{2.89}$$

Note that the boundary of each Q can be divided into two parts:

$$\sum_{e \in \partial Q} = \sum_{e \in \partial V} + \sum_{e \in \partial K},$$

where $\partial V(Q)$ and $\partial K(Q)$ are the finite volume and finite element edges associated with Q. Then the first terms in the right-hand sides of (2.88) and (2.89) can be formatted as follows:

$$\sum_{Q \subset \mathscr{Q}_H} \int_{\partial Q} \left(\frac{\partial u_H}{\partial n}(v - v_h^*) - p_H(v - v_h^*) \cdot n \right) ds$$

$$= \sum_{\partial Q} \left\{ \int_{\partial V} + \int_{\partial K} \right\} \left(\frac{\partial u_H}{\partial n}(v - v_h^*) - p_H(v - v_h^*) \cdot n \right) ds$$

$$= A(u_H, v_h^*) + D(v_h^*, p_H) + \sum_{E \in \mathscr{E}_H} \int_E J_E(v - v_H^*) ds$$

$$= (f, v_h^*) + \sum_{E \in \mathscr{E}_H} \int_E J_E(v - v_H^*) ds,$$

since

$$\sum_{Q \subset \mathscr{Q}_H} \int_{\partial V} \left(\frac{\partial u_H}{\partial n} - p_H \cdot n \right) v dx = 0.$$

Now, by using the second terms of (2.87)–(2.89), we obtain

$$(f, v - v_h^*) - \sum_{Q \subset \mathscr{Q}_H} \int_Q \left(-\Delta u_H(v - v_h^*) + \nabla p_H \cdot (v - v_h^*) \right) dx$$

$$= \sum_{Q \subset \mathscr{Q}_H} \int_Q (f + \Delta u_H - \nabla p_H) \cdot (v - v_h^*) dx$$

$$= \sum_{Q \subset \mathscr{Q}_H} \int_Q R_K \cdot (v - v_h^*) dx. \tag{2.90}$$

Finally, using all these equalities gives the desired result.   □

Theorem 2.6 *Let (u, p) and (u_H, p_H) be the solutions of Eqs. (2.4) and (2.27), respectively. Then we have*

$$\|(u - u_H, p - p_H)\|_\Omega \le C_5 \eta_H. \tag{2.91}$$

Proof For $w \in X$, there exists a $w_H \in X_H$ such that, for any $K \in \mathscr{T}_H$, $E \in \mathscr{E}_H$, $\omega_K = \bigcup_{K' \cap K \neq \emptyset} K'$, and $\omega_E = \bigcup_{K \cap E \neq \emptyset} K$ ([38, 115], the Scott–Zhang interpolation property),

$$\|w_H\|_{1,\Omega} \le C\|w\|_{1,\Omega},$$
$$\|w - w_H\|_{0,K} \le CH_K\|w\|_{1,\omega_K},$$
$$\|w - w_H\|_{0,E} \le CH_E^{1/2}\|w\|_{1,\omega_E}. \tag{2.92}$$

Using the Cauchy–Schwarz inequality in (2.86), and setting $(v, q) = (v - v_H, q)$ and $v_H^* = 0$, we see that

$$\left| \sum_{K \in \mathcal{T}_H} \int_K R_K \cdot (v - v_H) dx \right| \leq C \left(\sum_{K \in \mathcal{T}_H} H_K^2 \| R_K \|_{0,K}^2 \right)^{1/2} \| v \|_1,$$

$$\left| \sum_{K \in \mathcal{T}_H} \int_K \operatorname{div} u_H q dx \right| \leq C \sum_{K \in \mathcal{T}_H} \| \operatorname{div} u_H \|_{0,K} \| q \|_0,$$

$$\left| \sum_{E \in \mathcal{E}_H} \int_E J_E \cdot (v - v_H) ds \right| \leq C \left(\sum_{E \in \mathcal{E}_H} H_E \| J_E \|_{0,E}^2 \right)^{1/2} \| v \|_1.$$

Then it follows from the above inequalities and the continuous and coerciveness properties that

$$\begin{aligned}
&\| (u - u_H, p - p_H) \|_\Omega \\
&\leq \sup_{(v,q) \in X \times M} \beta^{-1} \frac{\mathcal{B}((u - u_H, p - p_H); (v, q))}{\| v \|_1 + \| q \|_0} \\
&\leq \beta^{-1} \left(\sum_{K \in \mathcal{T}_H} (\| H_K R_K \|_{0,K}^2 + \| \operatorname{div} u_H \|_{0,K}^2) + \sum_{E \in \mathcal{E}_H} H_E \| J_E \|_{0,E}^2 \right)^{1/2} \\
&\leq C_5 \eta_H,
\end{aligned}$$

which complete the proof. $\qquad\square$

2.6.2 Lower Bound

It remains to estimate a lower bound of the residual-based error estimator. Here, it is important to ensure the efficiency of an algorithm that uses η as a local refinement indicator, provided that a constant $C_6 > 0$ is determined to bound it below. For generation, we recall the definition of oscillation of a residual on each element $K \in \mathcal{T}_H$:

$$\begin{aligned}
osc_R^2(K) &\equiv \| H_K (R_K - \bar{R}_K) \|_{0,K}^2 \\
&= \| H_K (f - \bar{f}_K) \|_{0,K}^2
\end{aligned}$$

where we define $\bar{g}_K = \frac{1}{|K|} \int_K g dx$ for two terms $\bar{R}_K$ and $\bar{f}_K$. Moreover, we define the oscillation of the jump on each edge E by

$$osc_J^2(E) \equiv H_E \|J_E - \bar{J}_E\|_{0,E}^2$$
$$= H_E \|\frac{\partial u_H}{\partial n} - \frac{1}{H_E} \int_E \frac{\partial u_H}{\partial n} ds\|_{0,E}$$
$$= 0,$$

where $\bar{J}_E = \frac{1}{H_E} \int_E J_E ds$. Then the total oscillation is defined by

$$osc_H = \left(osc_R^2(K) + osc_J^2(E)\right)^{1/2}$$
$$= osc_R(K)$$

and the global oscillation by

$$osc_H = \left(\sum_{K \in \mathcal{T}_H} \|H_K(f - \bar{f}_K)\|_{0,K}^2\right)^{1/2}.$$

In order to analyze the residual based error estimator, the triangular bubble functions are introduced as follows: For a triangle $K \in \mathcal{T}_H$, let $\lambda_{K,1}$, $\lambda_{K,2}$ and $\lambda_{K,3}$ be the barycentric coordinates of K, and define the triangular bubble function b_K by

$$b_K(K) = \begin{cases} 27\lambda_{K,1}\lambda_{K,2}\lambda_{K,3} & \text{in } K, \\ 0 & \text{in } \Omega \backslash K. \end{cases} \tag{2.93}$$

Also, for any $E \in \mathcal{E}_H$, let the barycentric coordinates of the end points of E be $\lambda_{E,1}$ and $\lambda_{E,2}$, and define the edge bubble function b_E by

$$b_E(K) = \begin{cases} 4\lambda_{E,1}\lambda_{E,2} & \text{in } \omega_E, \\ 0 & \text{in } \Omega \backslash \omega_E. \end{cases} \tag{2.94}$$

Note that ω_E is defined in Theorem 2.6. Some useful results are recalled ([127], Sect. 4).

Lemma 2.7 *For any triangle $K \in \mathcal{T}_H$ and any edge $E \in \mathcal{E}_H$, the functions b_K and b_E satisfy the following properties:*

(1) supp $b_K \subset K$, $b_K \in [0,1]$, and $\max\limits_{x \in K} b_K(x) = 1$; supp $b_E \subset \omega_E$, $b_E \in [0,1]$, and $\max\limits_{x \in \omega_E} b_E(x) = 1$;

(2) $\int_K b_K dx = \frac{9}{20}|K| \sim H_K^2$, $\int_E b_E ds = \frac{2}{3} H_E$ and $\int_{\omega_E} b_E dx = \frac{1}{3}|\omega_E| \sim H_E^2$;

(3) $\|\nabla b_K\|_{0,K} \leq C H_K^{-1} \|b_K\|_{0,K}$ and $\|\nabla b_E\|_{0,\omega_E} \leq C H_E^{-1} \|b_E\|_{0,\omega_E}$.

Lemma 2.8 *For any $K \in \mathscr{T}_H$ and $(u_H, p_H) \in X_H \times M_H$, we have*

$$\|H_K \bar{R}_K\|_{0,K} \leq C \left(\|\!|(u - u_H, p - p_H)|\!\|_K + \|H_K(f - \bar{f}_K)\|_{0,K} \right),$$
$$\|H_K R_K\|_{0,K} \leq C \left(\|\!|(u - u_H, p - p_H)|\!\|_K + \|H_K(f - \bar{f}_K)\|_{0,K} \right). \quad (2.95)$$

Proof A proof of the a posteriori error estimator for second-order elliptic problems can be found in [132]. Here we show a lower bound of the residual-based error estimator for the Stokes problem. Obviously, it follows from Lemma 2.7 that

$$
\begin{aligned}
\|\bar{R}_K\|_{0,K}^2 &\sim (\bar{R}_K, b_K \bar{R}_K) \\
&= (R_K, b_K \bar{R}_K) - (R_K - \bar{R}_K, b_K \bar{R}_K) \\
&= a(u - u_H, b_K \bar{R}_K) - d(b_K \bar{R}_K, p - p_H) - (R_K - \bar{R}_K, b_K \bar{R}_K) \\
&\leq C \|\!|(u - u_H, p - p_H)|\!\|_K \|\nabla b_K \bar{R}_K\|_{0,K} + \|R_K - \bar{R}_K\|_{0,K} \|\bar{R}_K\|_{0,K}.
\end{aligned}
$$

From an inverse inequality and Lemma 2.7, it follows that

$$\|\nabla b_K \bar{R}_K\|_{0,K} \leq C H_K^{-1} \|\bar{R}_K\|_{0,K},$$
$$\|b_K \bar{R}_K\|_{0,K} \leq \|\bar{R}_K\|_{0,K}.$$

Thus, we have

$$\|\bar{R}_K\|_{0,K} \leq C H_K^{-1} \left(\|\!|(u - u_H, p - p_H)|\!\|_K + \|R_K - \bar{R}_K\|_{0,K} \right),$$

which implies the first inequality in (2.95). Also, it follows from the first inequality and the triangle inequality that

$$
\begin{aligned}
\|H_K R_K\|_{0,K} &\sim \|H_K(R_K - \bar{R}_K)\|_{0,K} + \|H_K \bar{R}_K\|_{0,K} \\
&\leq C \left(\|\!|(u - u_H, p - p_H)|\!\|_K + \|H_K(R_K - \bar{R}_K)\|_{0,K} \right);
\end{aligned}
$$

i.e., the second inequality in (2.95) holds. $\qquad\square$

Lemma 2.9 *For each edge $E \in \mathscr{E}_H$ and $(u_H, p_H) \in X_H \times M_H$, we have*

$$\|H_E^{1/2} \bar{J}_E\|_{0,E} \leq C \left(\|H_K R_K\|_{0,\omega_E} + \|\!|(u - u_H, p - p_H)|\!\|_{\omega_E} \right),$$
$$\|H_E^{1/2} J_E\|_{0,E} \leq C \left(\|H_K R_K\|_{0,\omega_E} + \|\!|(u - u_H, p - p_H)|\!\|_{\omega_E} \right). \quad (2.96)$$

Proof Similar to the proof of Lemma 2.8, using the properties of b_E and the Scott–Zhang interpolating property (2.92) and noting that [132]

$$|a(u - u_H, b_E \bar{J}_E) - d(b_E \bar{J}_E, p - p_H)|$$
$$\leq C H_E^{-1/2} \|(u - u_H, p - p_H)\|_{\omega_E} \|\bar{J}_E\|_{0,E},$$
$$|(R_K, b_E \bar{J}_E)_{\omega_E}| \leq C H_E^{1/2} \|R_K\|_{0,\omega_E} \|\bar{J}_E\|_{0,E}, \tag{2.97}$$

we obtain

$$\|\bar{J}_E\|_{0,E}^2 \sim (\bar{J}_E, b_E \bar{J}_E)_E$$
$$= (J_E, b_E \bar{J}_E)_E + (\bar{J}_E - J_E, b_E \bar{J}_E)_E$$
$$= a(u_H - u, b_E \bar{J}_E) - d(b_E \bar{J}_E, p_H - p) + (R_K, b_E \bar{J}_E)_{\omega_E}$$
$$\leq C \left(H_E^{-1/2} \|(u - u_H, p - p_H)\|_{\omega_E} + H_E^{1/2} \|R_K\|_{0,\omega_E} \right) \|\bar{J}_E\|_{0,E}.$$

Consequently, we have

$$\|H_E^{1/2} \bar{J}_E\|_{0,E} \leq C \left(\|(u - u_H, p - p_H)\|_{\omega_E} + \|H_K R_K\|_{0,\omega_E} \right).$$

Also, the second inequality in (2.96) follows from the triangle inequality and the first inequality. □

Now, we are ready to prove the main result by combining Lemmas 2.8 and 2.9.

Theorem 2.7 *Let $(u, p) \in X \times M$ and $(u_H, p_H) \in X_H \times M_H$ be the solutions of the Eqs. (2.4) and (2.27), respectively. Then we have*

$$\eta_H \leq C_6 \left(\|(u - u_H, p - p_H)\|_{\Omega} + osc_H \right). \tag{2.98}$$

Proof Using Lemmas 2.8 and 2.9 and taking a sum over all elements and edges yield the desired result. □

2.7   Adaptive Mixed Finite Volume Methods

The adaptive finite element method is an efficient and reliable algorithm for the numerical solution of enormous engineering problems. It is based on a posteriori error estimation for the finite element method, and the a posteriori error estimation involves easily computable local errors and serves as indication to provide information for adaptive local mesh refinement or mesh coarsening. Based on local error indicators, this method for solving partial differential equations includes successive loops of the following sequence:

$$\textbf{Solve} \rightarrow \textbf{Estimate} \rightarrow \textbf{Mark} \rightarrow \textbf{Refine}.$$

We refer to [2, 47, 127] and the references therein for more details.

There have been considerable efforts in the literature devoted to the development of efficient adaptive algorithms for solving various partial differential equations. So far, a rigorous convergence analysis of the above sequence relying on appropriate error reduction properties has been obtained for the conforming and nonconforming finite element methods [8, 23, 43, 79, 103–106, 126]. An optimal complexity analysis of the adaptive finite element methods can be found for second-order elliptic problems in [9, 121]. Optimal complexity for an adaptive method occurs if the exact solution of a differential equation is approximated by this method at an polynomial rate in the number of unknowns. In general, an iteratively constructed sequence of meshes can achieve this rate up to a constant factor.

Recently, the adaptive finite volume method has become very popular among the engineering community for flow computations. The main reason is that it not only has the prominent features of the finite element method in handling complex boundaries but also has the local conservation property of the finite difference method. In this chapter, we focus on an analysis of the adaptive stabilized mixed finite volume methods for the incompressible flow approximated by using the lower order elements. An abstract theory exists for a class of saddle-point problems [15, 59, 122]. The theory shows that their numerical approximations optimally converge if the finite element spaces for velocity and pressure satisfy a discretely uniform (with respect to a mesh step size) inf-sup condition (also called the Babuska–Brezzi condition). Nevertheless, the lower order finite element pairs do not satisfy this condition but are of practical importance in real applications. In particular, they are efficient for the saddle-point problem in terms of parallel and multigrid implementation because of the same mesh partition.

2.7.1 Discrete Local Lower Bound

In this subsection, we study a discrete local lower bound for $\||(u_h - u_H, p_h - p_H)\||$ between two conforming triangulations $\mathscr{T}_h$ and $\mathscr{T}_H$ and the corresponding finite element spaces $X_H \times M_H \subset X_h \times M_h$. Furthermore, an interior node property holds on each edge E of $\mathscr{T}_H$: The interior of $E = \partial \mathscr{T}_1 \cap \partial \mathscr{T}_2 (\mathscr{T}_1, \mathscr{T}_2 \in \mathscr{T}_H)$ contains at least one vertex of $\mathscr{T}_h$ [105, 132]. Compared with the results of the residual-based error estimator in the previous section, we focus on the discrete local lower bound on the domain ω_E for each $E \in \mathscr{E}_H$.

The main result in this section can be presented as follows:

Theorem 2.8 *Under the assumption of the interior node property, let $(u_h, p_h) \in X_h \times M_h$ and $(u_H, p_H) \in X_H \times M_H$ be the finite volume approximations of (2.4) on $\mathcal{T}_h$ and $\mathcal{T}_H$, respectively. Then we have*

$$\eta_H \leq C_7 \left(\| (u_h - u_H, p_h - p_H) \|_{\omega_E} + osc_H \right). \qquad (2.99)$$

In order to prove this theorem, we recall a relationship between the finite element and finite volume methods for the Stokes equations and apply similar ideas as in Lemmas 2.8 and 2.9.

Lemma 2.10 *Under the assumption of the interior node property, let $(u_h, p_h) \in X_h \times M_h$ and $(u_H, p_H) \in X_H \times M_H$ be the finite volume approximations of (2.4) on $\mathcal{T}_h$ and $\mathcal{T}_H$, respectively. Then we have, for $K_i \in \omega_E$, $i = 1, 2$,*

$$\| H_{K_i} R_{K_i} \|_{0,K_i} \leq C \left(\| (u_h - u_H, p_h - p_H) \|_{K_i} + \| f - \bar{f}_{K_i} \|_{0,K_i}^2 \right). \quad (2.100)$$

Proof The triangle inequality gives

$$\| R_{K_i} \|_{0,K_i} \leq \| R_{K_i} - \bar{R}_{K_i} \|_{0,K_i} + \| \bar{R}_{K_i} \|_{0,K_i}. \qquad (2.101)$$

For the second term in the right-hand side of (2.101), setting $\phi_i = I_h b_{K_i}$ and $\phi_i^* = I_h^* \phi_i \in X_h^*$ and applying the equivalence Lemma 2.3, the relationship between the finite element methods and the finite volume methods, an inverse inequality, and the interior node property, we find that

$$\begin{aligned}
\| \bar{R}_{K_i} \|_{0,K_i}^2 &\sim (\bar{R}_{K_i}, \phi_i^* \bar{R}_{K_i}) \\
&= (f + \Delta u_H - \nabla p_H, \phi_i^* \bar{R}_{K_i}) - (R_{K_i} - \bar{R}_{K_i}, \phi_i^* \bar{R}_{K_i}) \\
&= A(u_h - u_H, \phi_i^* \bar{R}_{K_i}) - D(\phi_i^* \bar{R}_{K_i}, p_h - p_H) - (R_{K_i} - \bar{R}_{K_i}, \phi_i^* \bar{R}_{K_i}) \\
&= a(u_h - u_H, b_K \bar{R}_{K_i}) - d(b_K \bar{R}_{K_i}, p_h - p_H) - (R_{K_i} - \bar{R}_{K_i}, \phi_i^* \bar{R}_{K_i}) \\
&\leq C \left(H_{K_i}^{-1} \| (u_h - u_H, p_h - p_H) \|_{K_i} + \| R_{K_i} - \bar{R}_{K_i} \|_{0,K_i} \right) \| \bar{R}_{K_i} \|_{0,K_i}. \quad (2.102)
\end{aligned}$$

Therefore, by (2.101) and (2.102),

$$\begin{aligned}
\| R_{K_i} \|_{0,K_i} &\leq C \left(\| R_{K_i} - \bar{R}_{K_i} \|_{0,K_i} + H_{K_i}^{-1} \| (u_h - u_H, p_h - p_H) \|_{K_i} \right) \\
&= C \left(\| f - \bar{f}_{K_i} \|_{0,K_i} + H_{K_i}^{-1} \| (u_h - u_H, p_h - p_H) \|_{K_i} \right). \quad (2.103)
\end{aligned}$$

Multiplying (2.103) by H_{K_i}, a straightforward computation shows (2.100).   □

Lemma 2.11 *Under the assumption of the interior node property, let $(u_h, p_h) \in X_h \times M_h$ and $(u_H, p_H) \in X_H \times M_H$ be the finite volume approximations of (2.4) on $\mathscr{T}_h$ and $\mathscr{T}_H$, respectively. Then we have, for $E \in \mathscr{E}_H$,*

$$\|H_E^{1/2} J_E\|_{0,E} \leq C \left(\|(u_h - u_H, p_h - p_H)\|_{\omega_E} + \|H_{K_i} R\|_{0,\omega_E} \right). \quad (2.104)$$

Proof Setting $\phi_E = I_h b_E$ and $\phi_E^* = I_h^* \phi_E$, we have

$$\|J_E\|_{0,E}^2 \sim (J_E, J_E \phi_E^*)$$

$$= \left(\frac{\partial u_H}{\partial n} - p_H I \cdot n, J_E \phi_E^* \right)_E. \quad (2.105)$$

Using the Green formula and the divergence theorem on each $Q \in \mathscr{Q}$ yields

$$\left(\frac{\partial u_H}{\partial n}, \phi_E^* \right)_E = -\int_{\partial V} \frac{\partial u_H}{\partial n} \phi_E^* ds + \int_Q \Delta u_H \cdot \phi_E^* dx,$$

$$(p_H I \cdot n, \phi_E^*)_E = -\int_{\partial V} \phi_E^* \cdot n p_H ds + \int_Q \nabla p_H \cdot \phi_E^* dx.$$

Thus, using these equalities and recalling the definition of $A(\cdot, \cdot)$ and $D(\cdot, \cdot)$, we see that

$$\left(\frac{\partial u_H}{\partial n} - p_H I \cdot n, \phi_E^* \right)_E$$

$$= -\int_{\partial V} \left(\frac{\partial u_H}{\partial n} - p_H I \cdot n \right) \phi_E^* ds + \int_Q (\Delta u_H - \nabla p_H) \cdot \phi_E^* dx,$$

$$= \int_{\partial V} \left(\frac{\partial (u_h - u_H)}{\partial n} - (p_h - p_H) I \cdot n \right) \phi_E^* ds + \int_Q (f + \Delta u_H - \nabla p_H) \cdot \phi_E^* dx.$$

Since $\text{supp} b_E \subset \omega_E$, multiplying J_E to both sides of the above equality, we have

$$\left| \int_{\partial V} \frac{\partial (u_h - u_H)}{\partial n} J_E \phi_E^* ds \right| = \left| A(u_h - u_H, J_E \phi_E^*) \right|$$

$$= \left| a(u_h - u_H, J_E \phi_E) \right|$$

$$\leq C H_E^{-1/2} \| u_h - u_H \|_{1, \omega_E} \| J_E \|_{0, E},$$

$$\left| \int_{\partial V} (p_h - p_H) I \cdot n J_E \phi_E^* ds \right| = \left| D(J_E \phi_E^*, p_h - p_H) \right|$$

$$= \left| d(J_E \phi_E, p_h - p_H) \right|$$

$$\leq C H_E^{-1/2} \| p_h - p_H \|_{0, \omega_E} \| J_E \|_{0, E},$$

$$\left| \int_Q (f + \Delta u_H - \nabla p_H) \cdot J_E \phi_E^* dx \right| \leq C H_E^{1/2} \| R_i \|_{0, \omega_E} \| J_E \|_{0, E}. \qquad (2.106)$$

Now, combining (2.105) and (2.106) yields

$$\| H_E^{1/2} J_E \|_{0, E} \leq C (\| (u_h - u_H, p_h - p_H) \|_{\omega_E} + \| H_{K_i} R_i \|_{0, \omega_E}).$$

$\square$

Finally, Theorem 2.8 follows from Lemmas 2.10 and 2.11.

2.7.2 Adaptive Finite Volume Algorithms

In this subsection, we present the adaptive finite volume methods by using the local error estimator presented in the previous sections. The techniques are adapted from [132] for second-order elliptic problems. For the Stokes equations, the adaptive finite volume methods can be divided into several steps. For the sake of convenience, we set $\{\mathcal{T}_k\}$, $k = 0, 1, 2 \ldots$, to be a sequence of shape-regular triangulations and $\{(u_k, p_k)\}$, $k = 0, 1, 2, \ldots$, to be a sequence of finite volume solutions on the corresponding nested finite element (volume) spaces generated by the adaptive finite volume methods. We consider the quasi-residual type of a posteriori error estimator and established the upper and (discrete local) lower bounds between the exact solution and the finite volume solutions and between the approximate solutions on successive meshes H and h, respectively.

First, we select parameters $\theta_1, \theta_2 \in (0, 1)$ and an initial partition $\mathcal{T}_0$ with the mesh scale $h_0 = H$.

Step I. Solve and estimate:
(1) Solve for the finite volume solutions (u_k, p_k):

$$\mathcal{C}_h((u_k, p_k); (v_k, q_k)) = (f, v_k^*).$$

(2) Compute the residual error estimator η_{h_k}.

Step II. Local refinement:
(1) Determine a suitable adaptive refinement for an update: Let $\mathscr{M}_k$ be the minimal edge set required refinement by satisfying the following marking strategy:

$$\theta_1 \eta_{h_k} \leq \eta_{h_k}(\mathscr{M}_k),$$
$$\theta_2 osc_{h_k} \leq osc_{h_k}(\mathscr{M}_k).$$

(2) Refinement and completion: Let $\mathscr{T}_{k+1}$ be the refinement of $\mathscr{T}_k$; refine each element $K \in \mathscr{M}_k$ and complete the hanging points such that $\mathscr{T}_{k+1}$ is a conforming triangulation.

Step III. Cycle criterion: For a sufficiently small tolerance $\varepsilon > 0$, if $\eta_k \leq \varepsilon$, then stop. Otherwise, set $k := k + 1$ and then go to step II.

Below a convergence analysis is given. It is known that the convergence analysis for the finite element method requires the orthogonality property [132]. However, the corresponding property loses effectiveness for the finite volume methods because the test and trial functions are different for the presented methods.

For convenience, set $(e_h, \varepsilon_h) = (u_h - u_H, p_h - p_H)$.

Lemma 2.12 *Let $(u_h, p_h) \in X_h \times M_h$ and $(u_H, p_H) \in X_H \times M_H \subset X_h \times M_h$ be the finite volume approximations of (2.4) on successive partitions $\mathscr{T}_h$ and $\mathscr{T}_H$, respectively. Then we have*

$$\mathscr{B}((u - u_h, p - p_h); (e_h, \varepsilon_h)) \leq C_8 \|\|(e_h, \varepsilon_h)\|\|_\Omega osc_h. \qquad (2.107)$$

Proof For the finite volume method, the two finite dimensional spaces X_h and X_h^* have the same dimension. For any $v_h \in X_h$ and $v_h^* = \Gamma_h v_h \in X_h^*$, there hold, for each interior element $K \in \mathscr{T}_h$ and its boundary ∂K [131, 136],

$$\int_K (v_h - v_h^*) dx = 0, \quad \int_{\partial K} (v_h - v_h^*) ds = 0,$$

satisfying Lemma 2.2. It then follows from (2.2) and Lemma 2.6 that

$$\int_K R_K \cdot (v_h - v_h^*) dx = \int_K (R_K - \bar{R}_K) \cdot (v_h - v_h^*) dx$$
$$\leq C h_K \| R_K - \bar{R}_K \|_{0,K} \| v_h \|_{1,K},$$
$$\int_K \operatorname{div} u_h q_h dx = \int_K (\operatorname{div} u_h - \bar{\operatorname{div}} u_h) q_h dx = 0,$$
$$\int_E J_E \cdot (v_h - v_h^*) ds = \int_E (J_E - \bar{J}_E)(v_h - v_h^*) ds = 0, \quad E \in \partial K.$$

Thus, summing up over all $K \in \mathcal{T}_h$ and $E \in \mathcal{E}_h$ with $(v_h, q_h) = (e_h, \varepsilon_h)$ and using the definition of osc_h yield the desired result. □

2.7.3 Convergence Analysis

The convergence property guarantees that the iterative loop terminates in a finite number of steps starting from an initial coarse mesh. Now, we investigate the adaptive finite volume methods for the Stokes equations in terms of an error reduction between two successive steps. Then the mathematical induction argument can be used to obtain the error reduction in the finite number of steps.

Lemma 2.13 *Let (u_h, p_h) and (u_H, p_H) be the finite volume solutions to (2.4) on the meshes $\mathcal{T}_h$ and $\mathcal{T}_H$, respectively, and let $H = \max\limits_{K \in \mathcal{T}_H} H_K$. Then there exists a positive constant $\rho_1 \in (0, 1)$ such that*

$$osc_h^2 \le \rho_1 osc_H^2. \tag{2.108}$$

Proof As example, we only estimate the term R_τ, $\tau \in \mathcal{T}_h$. Others can be bounded with a similar argument. For the residual on $\tau \subset K \in \mathcal{T}_H$, we have

$$R_\tau = R_\tau^H - \nabla(p_h - p_H), \tag{2.109}$$

where R_τ and R_τ^H are the error estimators on mesh $\mathcal{T}_h$ with solutions (u_h, p_h) and (u_H, p_H), respectively. Using the definition of R_τ, (2.109) and the Young inequality, we see that

$$
\begin{aligned}
osc_h^2 K &= \sum_{\tau \subset K} h_\tau^2 \| R_\tau - \bar{R}_\tau \|_{0,\tau}^2 \\
&\le \sum_{\tau \subset K} h_\tau^2 \| f - \bar{f} \|_{0,\tau}^2,
\end{aligned} \tag{2.110}
$$

since $\nabla \varepsilon_h$ is constant and thus $\nabla \varepsilon_h - \overline{\nabla \varepsilon_h} = 0$. Setting $\gamma_0 \in (0, 1)$ such that $h_\tau \le \gamma_0 H_K$ and $K \in \mathcal{M}_H$, we definite a refinement factor by

$$
\gamma_K = \begin{cases} \gamma_0 & \text{in } K \in \mathcal{M}_H, \\ 1 & \text{in } \Omega \backslash \mathcal{M}_H. \end{cases}
$$

By (2.16), (2.17), (2.110), the marking strategy on local refinement and straight-forward computations, then the above inequality restricted to the mesh $\mathcal{T}_H$ can be rewritten as follows:

$$\begin{aligned}
osc_h^2 &= \sum_{K \in \mathscr{T}_H} \sum_{\tau \subset K} osc_h^2(\tau) \\
&\leq \sum_{K \in \mathscr{T}_H} \gamma_K^2 osc_H^2(K) \\
&\leq \sum_{K \in \mathscr{T}_H \setminus \mathscr{M}_H} \gamma_0^2 osc_H^2(K) + \sum_{K \in \mathscr{M}_H} \gamma_0^2 osc_H^2(K) \\
&\leq osc_H^2 - (1 - \gamma_0^2) \sum_{K \in \mathscr{M}_H} osc_H^2(K) \\
&\leq (1 - (1 - \gamma_0^2)\theta_2^2) osc_H^2.
\end{aligned}$$ (2.111)

Using these above inequalities, we obtain

$$osc_h^2 \leq \rho_1 osc_H^2,$$

where $\rho_1 = (1 - (1 - \gamma_0^2))\theta_2^2 \in (0, 1)$. Therefore, (2.108) is shown.   □

Lemma 2.14 *Let (u_h, p_h) and (u_H, p_H) be the finite volume solutions to (2.4) on the meshes $\mathscr{T}_h$ and $\mathscr{T}_H$, respectively. There are constants $\gamma > 0$ and $\rho_0 \in (0, 1)$ such that*

$$\begin{aligned}
&\|(u - u_h, p - p_h)\|_\Omega^2 + \gamma osc_h^2 \\
&\leq \rho_0(\|(u - u_H, p - p_H)\|_\Omega^2 + \gamma osc_H^2).
\end{aligned}$$ (2.112)

Proof In view of $\|\cdot\|$ and the Young inequality, there exists a constant $\rho_2 \in (0, \frac{1}{2})$ such that

$$\begin{aligned}
\|(u - u_h, p - p_h)\|_\Omega^2 &\leq \|(u - u_H, p - p_H)\|_\Omega^2 - \|(u_h - u_H, p_h - p_H)\|_\Omega^2 \\
&\quad - \frac{1}{2}\|(u - u_h, p - p_h)\|_\Omega \|(u_h - u_H, p_h - p_H)\|_\Omega. \\
&\leq \|(u - u_H, p - p_H)\|_\Omega^2 - \|(u_h - u_H, p_h - p_H)\|_\Omega^2 \\
&\quad + \rho_2 \mathscr{B}((u - u_h, p - p_h); (u_h - u_H, p_h - p_H)).
\end{aligned}$$ (2.113)

Also, using Lemma 2.12 and the Young inequality gives

$$\begin{aligned}
&\rho_2 \mathscr{B}((u - u_h, p - p_h); (u_h - u_H, p_h - p_H)) \\
&\leq \rho_2 C_8 \|(u_h - u_H, p_h - p_H)\|_\Omega osc_h \\
&\leq \frac{\rho_2}{2}\|(u_h - u_H, p_h - p_H)\|_\Omega^2 + \frac{\rho_2 C_8^2}{2} osc_h^2,
\end{aligned}$$ (2.114)

which, together with (2.113) and Lemma 2.13 with respect to the relationship of the oscillations between $\mathscr{T}_h$ and $\mathscr{T}_H$, yields

$$
\begin{aligned}
&\|(u - u_h, p - p_h)\|_{\Omega}^2 + \gamma osc_h^2 \\
&\leq \|(u - u_H, p - p_H)\|_{\Omega}^2 - \frac{\rho_2}{2} \|(u_h - u_H, p_h - p_H)\|_{\Omega}^2 \\
&\quad + \beta_3 osc_H^2,
\end{aligned}
\tag{2.115}
$$

where $\beta_3 = (C_8^2 \rho_2 / 2 + \gamma)\rho_1$. By the results of the lower bound of the solution on the initial mesh and the discrete local lower bound, Theorems 2.6 and 2.8, and the marking strategy, we have

$$
\begin{aligned}
\|(u - u_H, p - p_H)\|_{\Omega}^2 &\leq C_5^2 \eta_H^2 \\
&\leq C_5^2 \theta_1^{-2} \eta_H^2(\mathscr{M}_H) \\
&\leq \beta_4 \left(\|(u_h - u_H, p_h - p_H)\|_{\Omega}^2 + osc_H^2 \right),
\end{aligned}
\tag{2.116}
$$

where $\beta_4 = C_5^2 C_7^2 \theta_1^{-2}$. Thus, we see that

$$
\begin{aligned}
&\|(u - u_h, p - p_h)\|_{\Omega}^2 + \gamma osc_h^2 \\
&\leq \left(1 - \frac{\rho_2}{2\beta_4} \right) \|(u - u_H, p - p_H)\|_{\Omega}^2 + \left(\beta_3 + \frac{\rho_2}{2} \right) osc_H^2.
\end{aligned}
$$

Choose appropriate parameters $\theta_1 > 0$, $\theta_2 > 0$ and $\gamma > 0$ such that

$$
1 - \frac{\rho_1}{2\beta_4} = 1 - \frac{\rho_1 \theta_1^2}{2C_5^2 C_7^2} \in (0, 1)
$$

and

$$
\beta_3 + \frac{\rho_2}{2} = \frac{1}{2} \left((C_8^2 \rho_2 + 2\gamma)(1 - (1 - \gamma^2))\theta_2^2 + \rho_2 \right) \in (0, 1).
$$

Therefore, setting $\rho_0 = \max\{1 - \frac{\rho_1}{2\beta_4}, \beta_3 + \frac{\rho_1}{2}\}$, the proof is completed. $\qquad\square$

Using induction for a series of partitions $\mathscr{T}_k$, $k = 0, 1, 2, \ldots$, we obtain the following convergence result:

Theorem 2.9 *Let $\{(u_k, p_k)\}$, $k = 0, 1, 2, \ldots$, be a sequence of the corresponding finite volume solutions on $\mathscr{T}_k$ defined above, and let the mesh size H of $\mathscr{T}_0$ be sufficiently small. Then we have*

$$\||(u - u_k, p - p_k)\||_{\Omega} \le \kappa_0 \rho_0^k, \tag{2.117}$$

where $\kappa_0 = \left(\||(u - u_H, p - p_H)\||_{\Omega}^2 + \gamma osc_H^2\right)^{1/2}$.

2.8 Numerical Experiments

This section presents numerical results that complement the theoretical analysis of Theorems 2.1. Our goal is to confirm the theoretical results of the stabilized finite volume methods for the two-dimensional stationary Stokes equations approximated by the lowest order finite element pairs. Three examples are considered: a nonphysical example with a known exact solutions, the well-known driven cavity problem and the backward-facing step problem.

In all experiments, the velocity is approximated by piecewise linear finite elements and the pressure is approximated by piecewise linear finite elements or piecewise constants defined with respect to the same uniform triangulation of Ω into triangles. In order to show the prominent features of these finite volume methods, the stabilized terms are computed in the following for the Stokes equations with a homogeneous or inhomogeneous boundary condition.

In order to filter unstable factors for the lowest equal order elements, we supply the local stabilized form of the difference between consistent and under-integrated mass matrices as follows [91]:

$$S_h(p_h, q_h) = p_i^T (M_k - M_1) q_j = p_i^T M_k q_j - p_i^T M_1 q_j, \tag{2.118}$$

for the implementation of the stabilized term Π_1. Here

$$p_i^T = [p_0, p_1, \ldots, p_{N_h-1}]^T, \quad q_j = [q_0, q_1, \ldots, q_{N_h-1}],$$

$$M_{ij} = (\phi_i, \phi_j), \quad p_h = \sum_{i=0}^{N_h-1} p_i \phi_i, \quad p_i = p_h(x_i), \quad \forall p_h \in M_h, \quad i, j = 0, 1, \ldots, N_h - 1,$$

ϕ_i is the basis function of pressure on the domain Ω such that its value is one at node x_i and zero at other nodes; the symmetric and positive M_k, $k \ge 2$ and M_1 are pressure mass matrices computed by using kth-order and first-order Gauss integration in each direction, respectively. Also, p_i and q_i, $i = 0, 1 \ldots, N_h - 1$, are the values of p_h and q_h at the node x_i. p_i^T is the transpose of the matrix p_i.

On the other hand, a suitable choice of Π_h^1 is to define its interpolation by using a projection onto the dual volume associated with each node [11]. For each node $p_j \in \mathcal{N}_h$, $j = 1, 2, \ldots, N_h$, the influence element $E_j \subset \Omega$ denotes the union of triangles that share the common vertex p_j, and ϕ_j denotes the continuous, piecewise linear basis function such that $\phi_j(p_m) = \delta_{j,m}$. We define the stabilized term $S_h(p_h, q_h) = (p_h - \Pi_h^0 p_h, p_h - \Pi_h^0 p_h)$ based on Clement-like interpolation:

$$\|p_h - \Pi_h^0 p_h\|_{0,K} = \left\|p_h - \sum_{p_j \in K} w_j \phi_j\right\|_{0,K}, \quad j = 0, 1, 2, \ldots, N_h - 1, \quad (2.119)$$

where w_j is computed by the following formula:

$$w_j = \frac{\sum\limits_{K \in E_j} V_j(K) p_K}{\sum\limits_{K \in E_j} V_j(K)}.$$

Here, p_K is the restriction of p_h on element $K \in \mathcal{T}_h$ and $V_j(K)$ is the volume of the element $K \in E_j$.

Therefore, it is easy to see that the discrete system of (2.27) is equivalent to a family of linear algebraic systems of the form

$$\begin{pmatrix} A & -D \\ D^T & G \end{pmatrix}$$

where the matrices A, D, and G are, respectively, deduced in the usual manner from the bilinear forms $a(\cdot, \cdot)$, $d(\cdot, \cdot)$, and $G(\cdot, \cdot)$. The computations of A, D, and D^T are preformed by the usual manner. In addition, the matrix G can be only accomplished by a little work at the element level.

Example 3.1 Let the computational domain be $\Omega = [0, 1] \times [0, 1]$ and $\mu = 1$. The boundary data and the source terms are chosen such that the exact solution of the Stokes equation is given by

$$\begin{cases} \mathbf{u} = [x^3 + x^2 y + x^2 - 3xy^2 - 2xy + x, -3x^2 y - xy^2 - 2xy + y^3 + y^2 - y]^T, \\ p = x^3 y^2 + xy + x + y - \frac{4}{3}. \end{cases}$$
$$(2.120)$$

In Table 2.1, we list the errors and estimated convergence rates (ECR) for a triangular mesh. The errors in the $\| \cdot \|$-norm are displayed for velocity and pressure of the Stokes equations. In all the cases, the numerical results obviously match the expected errors predicted by the theory (Table 2.2).

Also, numerical tests to check the L^∞ analysis for finite volume approximations of the Stokes problem were presented in [97]. We validate the theoretical convergence rates through a benchmark with the above analytical solution for the Stokes

Table 2.1 Stabilized FVMs: P_1-P_1

h	$\|\mathbf{u} - \mathbf{u}_h\|_0$	ECR	$\|\mathbf{u} - \mathbf{u}_h\|_1$	ECR	$\|p - p_h\|_0$	ECR
1/4	8.4314E-02		1.5127E-00		4.1190E-00	
1/8	2.3507E-02	1.8427	7.3035E-01	1.0505	1.8613E-00	1.1460
1/16	6.4762E-03	1.8599	3.5941E-01	1.0229	8.8411E-01	1.0740
1/32	1.7646E-03	1.8758	1.7840E-01	1.0105	4.3187E-01	1.0336
1/64	4.7560E-04	1.8915	8.8903E-02	1.0048	2.1376E-01	1.0146

Table 2.2 Stabilized FVMs: P_1-P_0

h	$\|\mathbf{u} - \mathbf{u}_h\|_0$	ECR	$\|\mathbf{u} - \mathbf{u}_h\|_1$	ECR	$\|p - p_h\|_0$	ECR
1/4	7.9555E-02		1.4687E-00		2.8851E-00	
1/8	2.2833E-02	1.8008	7.1538E-01	1.0377	1.2315E-00	1.2282
1/16	6.3430E-03	1.8479	3.5326E-01	1.0180	5.7111E-01	1.1086
1/32	1.7347E-03	1.8704	1.7566E-01	1.0080	2.7603E-01	1.0489
1/64	4.6867E-04	1.8881	8.7614E-02	1.0035	1.3592E-01	1.0221

equations approximated by using the macroelements iso-$P_2 - P_0$ and iso-$P_2 - P_1$ on two different triangulations with a box grid and a criss-cross grid in the previous paper ([97], Sect. 5. Numerical tests).

As seen from Figs. 2 and 3 in [97], the graphs here provide a more direct illustration of the datum and the slope well illustrates the convergence rates. Thus, the results here completely agree with theoretical expectations.

Example 3.2 The driven cavity is a standard test. It is a box full of liquid with its lid moving horizontally at speed one. In this example, the computational domain $\Omega = [0, 1] \times [0, 1]$ and $\mu = 1$. The boundary data is given by

$$\mathbf{u}_D = \begin{cases} [1, 0]^T & \text{if } y = 0, \\ [0, 0]^T & \text{else.} \end{cases} \tag{2.121}$$

Figure 2.2 shows the color contour of pressure p and streamlines of velocity $\mathbf{u}$ on a uniform mesh.

Example 3.3 Another benchmark problem is used to test the backward-facing step problem. Let the computational domain be $\Omega = [0, 4] \times [0, 1]\backslash[0, 1] \times [0, 0.5]$, and set $\mu = 1$. We assume that the Neumann boundary condition $\nabla\mathbf{u} \cdot \mathbf{n} = 0$ is imposed on the right boundary, and the Dirichlet boundary data is given by

$$\mathbf{u}_D = \begin{cases} [16(y - 0.5)(1 - y), 0]^T & \text{if } x = 0, \\ [0, 0]^T & \text{else.} \end{cases} \tag{2.122}$$

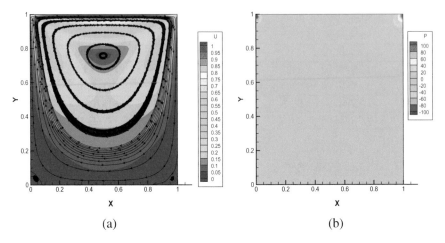

Fig. 2.2 Streamlines of velocity (left); Color contour of pressure (right)

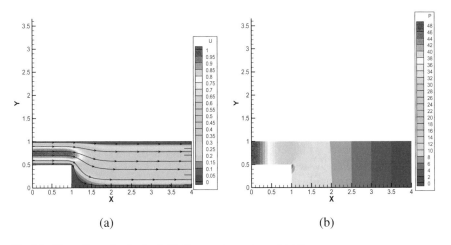

Fig. 2.3 Streamlines of velocity (left); Pressure isocontours (right)

This geometry produces a fluid recirculation zone that must be captured correctly. In this experiment, Fig. 2.3 displays the color contour of pressure p and streamlines of velocity $\mathbf{u}$ on a uniform mesh.

2.9 Conclusions

In this chapter, we consider the stabilized finite volume methods for the Stokes equations approximated by the lower order finite element pairs, which in fact can be easily applied for the other cases of lower order elements. After a relationship between these methods and the corresponding finite element methods is established, an error estimate of optimal order in the H^1-norm for velocity and an estimate in the L^2-norm for pressure are obtained with the same convergence rate as those provided by the corresponding finite element methods. Optimal error estimates for velocity in the L^2-norm and velocity gradient and pressure in the L^∞-norm are derived under an additional assumption on the body force.

The main results are derived in this section. For the finite volume solution (u_h, p_h), an error estimate of optimal order in the H^1-norm for velocity and an estimate in the L^2-norm for pressure are: If $f \in Y$, then

$$\|u - u_h\|_1 + \|p - p_h\|_0 \le \kappa h \left(\|u\|_2 + \|p\|_1 + \|f\|_0 \right),$$

where the constant κ is independent of the grid size h. If the body force satisfies $f \in [H^1(\Omega)]^d$, an optimal error estimate in the L^2-norm for velocity is

$$\|u - u_h\|_0 \le \kappa h^2 \left(\|u\|_2 + \|p\|_1 + \|f\|_1 \right).$$

The counterexample in [49, 74] showed that the finite volume solutions of elliptic problems approximated by the conforming linear elements cannot have an optimal L^2-norm convergence rate if the exact solution is in $H^2(\Omega)$ but the source term is only in $L^2(\Omega)$. Hence, based on the previous analysis, it is reasonable and optimal with additional regularity on the source term for the stationary Stokes equations.

In addition, some interesting relationships of the stabilized finite element methods and the corresponding finite volume methods can be found for the Stokes equations approximated by the lower order finite element pairs:

- The same linear algebraic matrices only for FEM and FVM;
- The same dimension for spaces of FEM and FVM;
- The same optimal estimate for the L^2-norm for velocity;
- The same optimal L^∞ estimates for velocity gradient and pressure;
- The superclose results between the solutions of FEM and FVM.

In detail, a super-close result is obtained for the solutions between the finite volume methods and the corresponding finite element methods with additional regularity on the body force $f \in [H^1(\Omega)]^d$:

$$\|\nabla(\bar{u}_h - u_h)\|_0 + \|\bar{p}_h - p_h\|_0 \le \kappa h^2 \left(\|u\|_2 + \|p\|_1 + \|f\|_1 \right).$$

Furthermore, the study is performed for the adaptive finite volume methods for these equations. First, we derive a residual type of a posteriori error estimator, and then establish both the upper bound

$$\|(u - u_H, p - p_H)\|_\Omega \le C_5 \eta_H,$$

and the lower bound

$$\eta_H \leq C_6(\|(u - u_H, p - p_H)\|_\Omega + osc_H),$$

in comparison with the exact error. Then, applying a technique developed in [105, 121, 132] and the relationship between the stabilized finite element methods and the stabilized finite volume methods, we establish a discrete local lower bound of the error between the finite volume solutions (u_H, p_H) and (u_h, p_h) on two successive meshes on $\mathscr{T}_H$ and $\mathscr{T}_h$ under the interior node property:

$$\eta_H \leq C_7(\|(u_h - u_H, p_h - p_H)\|_\Omega + osc_H).$$

For the adaptive finite volume methods, the loss of the Galerkin-orthogonality property is a major difficulty in analyzing their convergence for the Stokes equations. Under the assumption of a quasi-orthogonality property on the construction of dual partitions and using the same marking strategies as in [104, 105], the error reduction and convergence are obtained for the adaptive finite volume methods as follows:

$$\|(u - u_k, p - p_k)\|_\Omega \leq \kappa_0 \rho_0^k, \ 0 < \rho_0 < 1.$$

Obviously, the adaptive mixed finite volume methods converge with an exponential rate. Also, we can easily establish the optimal complexity of the presented methods for the incompressible flow. Furthermore, the optimal complexity of the adaptive mixed finite volume methods is determined by a negative exponential decay of the difference between the number of elements of the partition $\mathscr{T}_k$ and that of the initial partition $\mathscr{T}_0$.

Chapter 3
FVMs for the Stationary Navier–Stokes Equations

Abstract In this chapter, we mainly introduce the stabilized lower order finite volume methods for the stationary Navier–Stokes equations. Due to both the practical and theoretical needs, we will consider both small data and large data for the construction of the finite volume methods of the stationary Navier–Stokes equations.

3.1 Introduction

In this chapter, we extend the stabilized mixed finite element methods to the finite volume methods for the stationary Navier–Stokes equations. We will be concerned with the steady-state Navier–Stokes equations from the same point of view as in the previous chapter, i.e., existence, uniqueness and numerical approximation of a solution. However, some important differences from the previous linear case, such as nonlinearity, non-uniqueness, and theoretic analysis, arise.

While a finite volume method was studied for the Stokes equations [84, 96, 117, 135], the analysis for the Navier–Stokes equations needs special attentions due to the difficulties arising from the finite volume discretization for the nonlinear discrete terms. These nonlinear trilinear terms do not satisfy the anti-symmetry property any more. Thus, compared to the finite element analysis, the most difficulty here lies in the treatment of the nonlinear convection terms, which has a significant impact on theoretical analysis, especially on the L^2-norm estimate for velocity, L^∞-norm estimate for velocity gradient and pressure, and superclose result between the finite volume solution and the finite element solution.

The originality consists in extending systematically the results of the previous linear situation to the nonlinear situation. The basic result is a general theorem concerning the approximation and uniqueness of solutions (in general, only when "the data is small enough or the viscosity is large enough") and branches of nonsingular solutions for the stationary Navier–Stokes equations.

© The Author(s), under exclusive license to Springer Nature Switzerland AG 2022
J. Li et al., *Finite Volume Methods for the Incompressible Navier–Stokes Equations*,
SpringerBriefs in Mathematical Methods,
https://doi.org/10.1007/978-3-030-94636-4_3

In this chapter, existence and uniqueness of a solution to the nonlinear discrete Navier–Stokes equations is shown. Stability and convergence of optimal order in the H^1- and L^2-norm for velocity and pressure are obtained. Moreover, a new duality argument is introduced to establish the convergence of optimal order in the L^2 norm for velocity of the nonlinear Navier–Stokes flow [95]. An interesting result related to a superconvergence result is shown between the conforming mixed finite element solution and the finite volume solution using the same finite element pairs for the incompressible flow. Furthermore, optimal order estimates in the L^∞-norm for velocity gradient and pressure of this problem are obtained by using new technical results without the presence of a usual logarithmic factor $O(|\log h|)$.

This chapter is organized as follows: In the next section, we introduce finite volume approximations for the stationary Navier–Stokes equations with small data. Then, in the third section, a branch of nonsingular finite volume solutions of the stationary Navier–Stokes equations is considered. Finally, some conclusions are drawn in the last section.

In this chapter, more attention is paid to the investigation of numerical analysis for the stationary Navier–Stokes equations. The main idea of this chapter is to include both two- and three-dimensional problems in numerical analysis, and not limit itself to the two-dimensional case. Moreover, it is particularly important to compare the difference between the two- and three-dimensional scenarios.

3.2 FVMs for the Stationary Navier–Stokes with Small Data

The uniqueness of a solution of the stationary Navier–Stokes equations has only been proved under the assumption of small data (μ is sufficiently large, or the given forces and boundary values of velocity are sufficiently small). In this section, we will establish the corresponding existence and uniqueness theorems and optimal convergence of mixed finite volume solutions of the stationary Navier–Stokes problem with small data.

3.2.1 The Weak Formulation

We consider the stationary Navier–Stokes equations: With $\Omega \subset R^d, d = 2, 3$, being a bounded open set with the Lipschitz boundary $\partial\Omega$,

$$-\mu\Delta u + \nabla p + (u \cdot \nabla)u = f \quad \text{in } \Omega, \tag{3.1}$$

$$\operatorname{div} u = 0 \quad \text{in } \Omega, \tag{3.2}$$

$$u = 0 \quad \text{on } \partial\Omega, \tag{3.3}$$

where u represents the velocity vector, p the pressure, f the prescribed body force, and $\mu > 0$ the viscosity.

Using the bilinear forms defined in Chap. 2, the weak formulation associated with (3.1)–(3.3) is to seek $(u, p) \in X \times M$ such that

$$\mathscr{B}((u, p); (v, q)) + b(u, u, v) = (f, v) \quad \forall (v, q) \in X \times M, \tag{3.4}$$

where the trilinear form is

$$b(u, v, w) = ((u \cdot \nabla)v, w) + \frac{1}{2}((\operatorname{div} u)v, w) \tag{3.5}$$

$$= \frac{1}{2}((u \cdot \nabla)v, w) - \frac{1}{2}((u \cdot \nabla)w, v), \quad u, v, w \in X. \tag{3.6}$$

As mentioned above, a further assumption on Ω is needed:

Assumption (A1) Assume that Ω is regular in the sense that the unique solution (u, p) of the stationary Stokes problem for a prescribed $g \in [L^r(\Omega)]^d$ exists and satisfies

$$\|v\|_{2,r} + \|q\|_{1,r} \le C\|g\|_{L^r},$$

where $1 \le r \le \infty$.

Obviously, the validity of assumption (A1) is known if Γ is of C^2. Then, there holds the following inequalities [1, 80]:

$$\|v\|_0 \le \gamma\|v\|_1, \quad v \in X,$$
$$\|v\|_{L^4} \le 2^{\frac{d-1}{4}}\|v\|_0^{\frac{d}{4}}\|v\|_1^{1-\frac{d}{4}},$$
$$\|v\|_{L^\infty} \le C_9\|v\|_0^{1-\frac{d}{4}}\|v\|_2^{\frac{d}{4}}, \quad v \in D(A), \tag{3.7}$$

where the positive constant $\gamma > 0$ only depends on Ω, and the second and third inequalities can be found in ([80], Lemmas 1 and 2 for $d = 2, 3$) and ([137], Gagliardo-Nirenberg interpolation inequality), respectively.

Then the trilinear form $b(\cdot; \cdot, \cdot)$ satisfies [67, 70, 92, 122]:

$$b(u, v, w) = -b(u, w, v) \quad \forall u, v, w \in X, \tag{3.8}$$
$$|b(u, v, w)| \le C_{10}\|u\|_1 \|v\|_1 \|w\|_1 \quad \forall u, v, w \in X, \tag{3.9}$$
$$|b(u, v, w)| + |b(v, u, w)| + |b(w, u, v)| \le C_{10}\|u\|_1 \|Av\|_0\|w\|_0,$$
$$\forall u \in X, v \in D(A), w \in Y. \tag{3.10}$$

Detailed results on existence and uniqueness of a solution to Eqs. (3.1) and (3.2) are known ([122], Theorem 1.2). In particular, we state the next theorem.

Theorem 3.1 *If $\mu > 0$ and $f \in Y$ satisfy*

$$1 - \frac{C_{10}\gamma}{\mu^2} \|f\|_0 > 0, \tag{3.11}$$

then the variational problem (3.4) *admits a unique solution* $(u, p) \in D(A) \times H^1(\Omega) \cap M$ *satisfying*

$$\|u\|_1 \leq \frac{\gamma}{\mu}\|f\|_0, \quad \|u\|_2 + \|p\|_1 \leq C\|f\|_0, \tag{3.12}$$

where the positive constants γ and C_{10} are given by (3.7) *and* (3.9)*, respectively.*

3.2.2   Galerkin FV Approximation

After a quick description of the problem, we will present some basic concepts which can be used as an introduction to the stationary Navier–Stokes equations. This will be followed by a more detailed discussion. It will not be possible to analyze all the cases in detail; we shall try to group them by families which can be treated by similar methods. These families will be arbitrary and will overlap in many cases. Then we shall briefly discuss some modified variational formulations that can be used to obtain better results for the finite volume versions.

For the subsequent analysis, we now introduce a discrete analogue A_h of the Laplace operator A through the condition ([71], Sect. 6)

$$(A_h u_h, v_h) = (\nabla u_h, \nabla v_h), \ u_h, v_h \in X_h.$$

Define
$$V_h = \{v_h \in X_h : d(v_h, q_h) = 0 \ \forall \ q_h \in M_h\}.$$

The restriction of A_h to V_h is invertible, with the inverse A_h^{-1}. In addition, A_h is self-adjoint and positive definite. Especially, we define the discrete Sobolev norm on V_h for any $r \in R$ by

$$\|v_h\|_r = \|A_h^{r/2} v_h\|_{L^2}, \quad v_h \in V_h.$$

Based on the previous results, the corresponding discrete finite element variational formulation of (3.4) for the Navier–Stokes equations is recast: Find $(\bar{u}_h, \bar{p}_h) \in X_h \times M_h$ such that

$$\mathscr{B}_h((\bar{u}_h, \bar{p}_h), (v_h, q_h)) + b(\bar{u}_h, \bar{u}_h, v_h) = (f, v_h) \ \forall \ (v_h, q_h) \in X_h \times M_h. \tag{3.13}$$

Because of (2.19) and (2.20), the existence, uniqueness, and regularity of a solution to system (3.13) can be easily verified [59, 122]. Moreover, an optimal estimate for the stabilized finite element solution (u_h, p_h) holds (Theorem 4.3 in [67] and Theorem 4.2 in [69])

$$\|u - \bar{u}_h\|_0 + h\left(\|u - \bar{u}_h\|_1 + \|p - \bar{p}_h\|_0\right) \le \kappa h^2 \left(\|u\|_2 + \|p\|_1\right). \quad (3.14)$$

Then, the stabilized mixed finite volume methods for the Navier–Stokes equations (3.4) are: Find $(u_h, p_h) \in (X_h, M_h)$ such that

$$\mathscr{C}_h((u_h, p_h), (v_h, q_h)) + b(u_h, u_h, v_h^*) = (f, v_h^*) \quad \forall (v_h, q_h) \in X_h \times M_h. \quad (3.15)$$

Here, using a technique similar to the trilinear form of the finite element method in the previous section, we define the trilinear form $b(\cdot; \cdot, \cdot) : X_h \times X_h \times X_h^* \to \Re$ of the finite volume method [122]:

$$b(u_h, v_h, w_h^*) = \left((u_h \cdot \nabla)v_h + \frac{1}{2}(\operatorname{div} u_h)v_h, w_h^*\right) \quad \forall u_h, v_h, w_h \in X_h.$$

Note that the definition of $b(\cdot, \cdot, \cdot)$ of the finite volume methods remains consistent with the continuous case. A fundamental difference between it and that of the finite element method lies in the test and trial functions defined in two different spaces. As noted, the difficulty in the finite volume methods is that the trilinear term no longer satisfies the useful skew-symmetry property in the context of the Petrov–Galerkin method. Thus, stability and convergence of these methods are more difficult than those of the finite element methods for the stationary Navier–Stokes equations.

Now, the existence, uniqueness, and regularity are established in the theorem below.

3.2.3 Existence and Uniqueness Theorem

In order to get a well-posed problem for the stationary Navier–Stokes equations (3.15), we are now in the position of staring the main results of this subsection. For this, we define the mesh parameter

$$h_0(h) = \begin{cases} \dfrac{4C_2C_3C_4\gamma}{\mu^2} |\log h|^{1/2}h\|f\|_0, & d = 2, \\[3ex] \dfrac{4C_2C_3C_4\gamma}{\mu^2} h^{1/2}\|f\|_0, & d = 3. \end{cases}$$

Theorem 3.2 (Existence and uniqueness theorem) *For each $h > 0$ such that*

$$0 < h_0 \le 1/2, \tag{3.16}$$

system (3.15) *admits a solution* $(u_h, p_h) \in X_h \times M_h$. *Moreover, if the viscosity* $\mu > 0$, *the body force* $f \in Y$, *and the mesh size* $h > 0$ *satisfy*

$$0 < h_0 \le \frac{1}{4}, \quad 1 - \frac{4C_4 C_{10}\gamma}{\mu^2}\|f\|_0 > 0, \tag{3.17}$$

then the solution $(u_h, p_h) \in X_h \times M_h$ *is unique. Furthermore, it satisfies*

$$\|u_h\|_1 \le \frac{2C_4\gamma}{\mu}\|f\|_0,$$

$$\|p_h\|_0 \le \frac{2C_4\gamma\|f\|_0}{\mu^2\beta^*}(\mu^2 + 2\gamma C_4 C_{10}\|f\|_0),$$

$$\|A_h u_h\|_0^2 \le \frac{4C_4^2\|f\|_0^2}{\mu^2} + \frac{4C_d}{\mu}, \quad d = 2, 3, \tag{3.18}$$

where C_d is defined below.

Proof For fixed $f \in Y$, we introduce the set

$$B_h = \left\{ (v_h, q_h) \in X_h \times M_h : \|v_h\|_1 \le \frac{2C_4\gamma}{\mu}\|f\|_0, \right.$$

$$\left. \|p_h\|_0 \le \frac{2C_4\gamma\|f\|_0}{\mu^2\beta^*}(\mu^2 + 2C_4 C_{10}\gamma\|f\|_0) \right\}.$$

Then we define the mapping $T_h : B_h \to X_h \times M_h$ to obtain the following equations: Given $(\bar{w}_h, \bar{r}_h) \in B_h$, find $T_h(\bar{w}_h, \bar{r}_h) = (w_h, r_h)$ such that for all $v_h \in X_h$,

$$\mathscr{C}_h((w_h, r_h), (v_h, q_h)) + b(\bar{w}_h; w_h, v_h^*) = (f, v_h^*), \quad (v_h, q_h) \in X_h \times M_h. \tag{3.19}$$

Applying Brouwer's fixed point theorem, we will prove that T_h maps B_h into B_h.

First, taking $(v_h, q_h) = (w_h, r_h) \in X_h \times M_h$ in (3.19), we see that

$$\mathscr{C}_h((w_h, r_h), (w_h, r_h)) + b(\bar{w}_h, w_h, w_h^* - w_h) + b(\bar{w}_h, w_h, w_h) = (f, w_h^*). \tag{3.20}$$

By using the definition of $b(\cdot; \cdot, \cdot)$ and $\mathscr{C}_h(\cdot, \cdot)$, (2.25), (3.7), and the equivalence Lemma 2.3, we have

$$|\mathscr{C}_h((w_h, r_h), (w_h, r_h))| \ge \mu\|w_h\|_1^2,$$
$$|(f, w_h^*)| \le \|f\|_0\|w_h^*\|_0 \le \gamma C_4\|f\|_0\|w_h\|_1.$$

Thanks to (3.8),

$$|b(\bar{w}_h, w_h, w_h)| = 0.$$

Applying Hölder's inequality yields

$$|b(\bar{w}_h, w_h, w_h^* - w_h)| \leq \left(\|\bar{w}_h\|_{L^\infty} \|w_h\|_1 + \frac{\sqrt{d}}{2} \|\bar{w}_h\|_1 \|w_h\|_{L^\infty} \right) \|w_h^* - w_h\|_0$$

$$\leq 2C_3 h \left(\|\bar{w}_h\|_{L^\infty} \|w_h\|_1 + \frac{\sqrt{d}}{2} \|\bar{w}_h\|_1 \|w_h\|_{L^\infty} \right) \|w_h\|_1.$$

In the two-dimensional case, we use the inverse inequality (2.12) and the set B_h to obtain

$$|b(\bar{w}_h, w_h, w_h^* - w_h)| \leq 2C_2 C_3 |\log \frac{1}{h}|^{1/2} h \|\bar{w}_h\|_1 \|w_h\|_1^2$$

$$\leq \frac{4C_2 C_3 C_4 \gamma}{\mu} |\log \frac{1}{h}|^{1/2} h \|f\|_0 \|w_h\|_1^2.$$

Similarly, applying (2.13) yields the bound in three dimensions:

$$|b(\bar{w}_h, w_h, w_h^* - w_h)| \leq 2C_2 C_3 h^{1/2} \|\bar{w}_h\|_1 \|w_h\|_1^2$$

$$\leq \frac{4\gamma C_2 C_3 C_4}{\mu} h^{1/2} \|f\|_0 \|w_h\|_1^2.$$

Therefore, we have

$$|b(\bar{w}_h, w_h, w_h^* - w_h)| \leq \mu h_0 \|w_h\|_1^2, \tag{3.21}$$

which, together with these inequalities, gives

$$\mu (1 - h_0) \|w_h\|_1 \leq C_4 \gamma \|f\|_0. \tag{3.22}$$

Using (3.16), it holds

$$\|w_h\|_1 \leq \frac{2C_4 \gamma}{\mu} \|f\|_0. \tag{3.23}$$

In view of the Eq. (3.19), the weak coercivity (2.31) of $\mathscr{C}((\cdot, \cdot), (\cdot, \cdot))$, and (3.16), we find that

$$\|r_h\|_0 \leq \frac{1}{\beta^*} \sup_{(v_h, q_h) \in X_h \times M_h} \frac{\mathscr{C}_h((w_h, r_h), (v_h, q_h))}{\|v_h\|_1 + \|q_h\|_0}$$

$$\leq \frac{1}{\beta^*} (\mu h_0 \|w_h\|_1 + C_{10} \|v_h\|_1 \|w_h\|_1 + C_4 \gamma \|f\|_0)$$

$$\leq \frac{2C_4 \gamma \|f\|_0}{\mu^2 \beta^*} (\mu^2 + 2C_4 C_{10} \gamma \|f\|_0). \tag{3.24}$$

Since the mapping T_h is well defined, it follows from Brouwer's fixed point theorem that there exists a solution to system (3.15).

Then, we assume that (u_1, p_1) and (u_2, p_2) are two solutions to (3.15). Next, we have

$$\mathscr{C}_h((u_1 - u_2, p_1 - p_2), (v_h, q_h)) + b(u_1 - u_2, u_1, v_h^*)$$
$$+ b(u_2, u_1 - u_2, v_h^*) = 0. \tag{3.25}$$

By setting $(v_h, q_h) = (u_1 - u_2, p_1 - p_2) = (e, \eta)$, we can see that

$$\mathscr{C}_h((e, \eta), (e, \eta)) \geq \mu \|e\|_1^2. \tag{3.26}$$

By the inverse inequality (2.12), estimate (2.24) between two different systems, the bounds for $\|u_h\|_1$ and the trilinear term in (3.9) and (3.23), we see that

$$|b(e, u_1, e^*) + b(u_2; e, e^*)| = |b(e, u_1, e^* - e) + b(u_2, e, e^* - e) + b(e, u_1, e)|$$
$$\leq C_{10} \|u_1\|_1 \|e\|_1^2 + 2\mu h_0 \|e\|_1^2$$
$$\leq \left(\frac{2C_4 C_{10} \gamma}{\mu} \|f\|_0 + 2\mu h_0 \right) \|e\|_1^2. \tag{3.27}$$

Applying (3.25)–(3.27) and (3.17) gives

$$0 < \mu \left(1 - \frac{4C_4 C_{10} \gamma}{\mu^2} \|f\|_0 \right) \|e\|_1^2 \leq 0, \tag{3.28}$$

which shows that $e = 0$; i.e., $u_1 = u_2$. Next, applying (2.31) to (3.25) yields that $p_1 = p_2$. Therefore, it follows that (3.15) has a unique solution.

Finally, using the definition of the discrete operator A_h and taking $(v_h, q_h) = (A_h u_h, 0)$ in (3.15), it follows that

$$\mu \|A_h u_h\|_0^2 + b(u_h, u_h, (A_h u_h)^*) = (f, (A_h u_h)^*), \tag{3.29}$$

where, using Hölder's inequality,

$$|(f, (A_h u_h)^*)| \leq \|f\|_0 \|(A_h u_h)^*\|_0 \leq \frac{C_4^2}{\mu} \|f\|_0^2 + \frac{\mu}{4} \|A_h u_h\|_0^2.$$

Now, by employing the L^∞ estimate inequality (3.7) and (2.25), we conclude that

$$
\begin{aligned}
&|b(u_h, u_h, (A_h u_h)^*)| \\
&\leq C_4 C_9 \|u_h\|_0^{1/2} \|u_h\|_1 \|A_h u_h\|_0^{3/2} + \frac{\sqrt{d}}{2} C_{10} \|u_h\|_1 \|u_h\|_0^{1/2} \|A_h u_h\|_0^{3/2} \\
&\leq \frac{\mu}{4} \|\|\|_0^2 + C_d,
\end{aligned}
\tag{3.30}
$$

where

$$
C_d = \begin{cases}
\dfrac{2^4 C_4^4 C_9^4 \|u_h\|_1^6}{\mu^3}, & d = 2, \\
\dfrac{2^{11} C_4^8 C_9^8 \|u_h\|_1^5}{\mu^7}, & d = 3.
\end{cases}
\tag{3.31}
$$

which, together with (3.29), completes the proof of (3.18).   □

3.2.4   Convergence Analysis

The main result of this subsection consists in showing error estimates for the finite volume methods governed by the stationary Navier–Stokes equations. This subsection is organized in two aspects. In the first one, we discuss a superclose property between the finite element and finite volume solutions. The second one deals with optimal estimates of the finite volume methods.

3.2.4.1   Optimal and Superclose Properties

Optimal and superclose results are also obtained between the finite element and finite volume solutions. We will use an assumption stronger than (3.17):

$$
1 - \frac{2 C_4 C_{10} \gamma}{\mu^2} \|f\|_0 \geq C_{11} > 0.
\tag{3.32}
$$

Lemma 3.1 *Assume that $h > 0$ satisfies (3.16), and $f \in Y$ and $\mu > 0$ satisfy (3.32). Let $(\bar{u}_h, \bar{p}_h) \in X_h \times M_h$ and $(u_h, p_h) \in X_h \times M_h$ be the solutions of (3.13) and (3.15), respectively. If $f \in [H^i(\Omega)]^d$, then it holds*

$$
\|\bar{u}_h - u_h\|_1 + \|\bar{p}_h - p_h\|_0 \leq C h^{1+i/2} \|f\|_i, \quad i = 0, 1.
\tag{3.33}
$$

Proof Subtracting (3.15) from (3.13), it follows from the equivalence Lemma 2.3 that

$$\mathscr{C}_h((\bar{u}_h - u_h, \bar{p}_h - p_h), (v_h, q_h)) + b(\bar{u}_h, \bar{u}_h, v_h) - b(u_h, u_h, v_h^*)$$
$$= (f, v_h - v_h^*). \quad (3.34)$$

Taking $(v_h, q_h) = (e, \eta) = (\bar{u}_h - u_h, \bar{p}_h - p_h)$ in (3.34), we see that

$$\mathscr{C}_h((e, \eta), (e, \eta)) + b(e, \bar{u}_h, e) + b(u_h, e, e) + b(u_h, u_h, e - e^*)$$
$$= (f, e - e^*). \quad (3.35)$$

Clearly, using (3.8) leads to the following result:

$$b(u_h, e, e) = 0. \quad (3.36)$$

Thanks to the estimates in (3.9) and (3.18) for the trilinear term and the energy norm of $\|u_h\|_1$, we have

$$|b(e, u_h, e)| \le C_{10}\|e\|_1^2\|u_h\|_1 \le \frac{2C_4C_{10}\gamma}{\mu}\|f\|_0\|e\|_1^2. \quad (3.37)$$

Owing to the definition of the bilinear term $\mathscr{C}_h((\cdot, \cdot), (\cdot, \cdot))$ gives

$$|\mathscr{C}_h((e, \eta), (e, \eta))| \ge \mu\|e\|_1^2. \quad (3.38)$$

In addition, setting $\mu_0 = 1 - \frac{2C_4C_{10}\gamma\|f\|_0}{\mu^2}$ and using Lemma 2.2, the inverse inequality (2.13), and the same approach as for (2.43), we see that

$$|b(u_h, u_h, e - e^*)|$$
$$= \left|\left(((u_h - \hat{\pi}_h u_h) \cdot \nabla)u_h + \frac{1}{2}\text{div } u_h(u_h - \hat{\pi}_h u_h), e - e^*\right)\right|$$
$$\le \left\{\|A_h^{1/2}u_h\|_{L^\infty}\|u_h - \hat{\pi}_h u_h\|_0 + \frac{\sqrt{d}}{2}\|A_h^{1/2}u_h\|_{L^\infty}\|u_h - \hat{\pi}_h u_h\|_0\right\}\|e - e^*\|_0$$
$$\le Ch^{3/2}\|A_h u_h\|_0\|u_h\|_1\|e\|_1$$
$$\le Ch^3\|u_h\|_1^2\|A_h u_h\|_0^2 + \frac{\mu_0}{4}\|e\|_1^2. \quad (3.39)$$

Similarly, using the same approach as for (2.43), we have

$$|(f, e - e^*)| = |(f - \hat{\pi}_h f, e - e^*)|$$
$$\le Ch^{1+i}\|f\|_i\|e\|_1 \le Ch^{2(1+i)}\|f\|_i^2 + \frac{\mu_0}{4}\|e\|_1^2. \quad (3.40)$$

Combining (3.17) with (3.35)–(3.40) gives

$$\|e\|_1 \le Ch^{1+i/2}\|f\|_i. \quad (3.41)$$

In the same argument, it follows from (2.31) and (3.34) that

$$\|\eta\|_0 \leq Ch^{1+i/2}\|f\|_i. \tag{3.42}$$

Combining (3.41) with (3.42) completes the proof of (3.33). □

Remark From (3.33), it shows that the convergence order of the superclose result for the stationary Navier–Stokes equations seems to be a bit lower than that for the stationery Stokes equations since the bound of the trilinear term has a negative effect on the convergence analysis.

Remark Furthermore, using the first inverse inequality in (2.13), we can obtain a more accurate estimate of Theorem 3.1 of this problem in two dimensions:

$$\|\bar{u}_h - u_h\|_1 + \|\bar{p}_h - p_h\|_0 \leq \kappa |\log \frac{1}{h}|h^2(\|u\|_2 + \|p\|_1 + \|f\|_1), \tag{3.43}$$

where the factor $|\log \frac{1}{h}|^{1/2}$ is more accurate than the factor $O(h^{1/2})$.

Thus, we can obtain the following superclose property:

Theorem 3.3 (Superclose-property) *Under the assumption of Theorem 3.2, let $(\bar{u}_h, \bar{p}_h) \in X_h \times M_h$ and $(u_h, p_h) \in X_h \times M_h$ be the solutions of (3.13) and (3.15), respectively. If $f \in [H^1(\Omega)]^d$, then it holds for two dimensions*

$$\|\bar{u}_h - u_h\|_1 + \|\bar{p}_h - p_h\|_0 \leq \kappa |\log \frac{1}{h}|h^2(\|u\|_2 + \|p\|_1 + \|f\|_1) \tag{3.44}$$

and for three dimensions

$$\|\bar{u}_h - u_h\|_1 + \|\bar{p}_h - p_h\|_0 \leq \kappa h^{3/2}(\|u\|_2 + \|p\|_1 + \|f\|_1). \tag{3.45}$$

3.2.4.2 Optimal Analysis

Based on the above results, we can obtain the following optimal estimates of velocity in the L^2- and H^1-norms, and pressure in the L^2-norm for the stationary Navier–Stokes equations:

Theorem 3.4 *Assume that $h > 0$ satisfies (3.16), and $f \in Y$ and $\mu > 0$ satisfy (3.32). Let $(u, p) \in X \times M$ and $(u_h, p_h) \in X_h \times M_h$ be the solutions of (3.4) and (3.15), respectively. Then it holds*

$$\|u - u_h\|_1 + \|p - p_h\|_0 \leq \kappa h(\|u\|_2 + \|p\|_1 + \|f\|_0). \tag{3.46}$$

Proof Using a triangle inequality, (3.14), (3.41), and (3.42), we obtain

$$\|u - u_h\|_1 \leq \|u - \bar{u}_h\|_1 + \|\bar{u}_h - u_h\|_1 \leq Ch(\|u\|_2 + \|p\|_1 + \|f\|_0),$$
$$\|p - p_h\|_0 \leq \|p - \bar{p}_h\|_0 + \|\bar{p}_h - p_h\|_0 \leq Ch(\|u\|_2 + \|p\|_1 + \|f\|_0),$$

which completes the proof of (3.46). $\qquad\qquad\qquad\qquad\qquad\qquad\qquad\qquad\square$

The L^2 estimate for velocity is the most difficult in error estimates of the present problem since the Petrov–Galerkin system loses symmetry. In order to derive an optimal error estimate for velocity in the L^2-norm, we consider the following dual problem:

$$a(v, \Phi) + d(v, \Psi) - d(\Phi, q) + b(u, v, \Phi) + b(v, u, \Phi) = (u - u_h, v). \tag{3.47}$$

Because of convexity of the domain Ω, this problem has a unique solution that satisfies the regularity property [122]

$$\|\Phi\|_2 + \|\Psi\|_1 \leq C\|u - u_h\|_0. \tag{3.48}$$

Below set $(\Phi_h, \Psi_h) = (I_h\Phi, J_h\Psi) \in X_h \times M_h$, which satisfies, by (2.10),

$$\|\Phi - \Phi_h\|_0 + h(\|\Phi - \Phi_h\|_1 + \|\Psi - \Psi_h\|_0) \leq Ch^2(\|\Phi\|_2 + \|\Psi\|_1). \tag{3.49}$$

Theorem 3.5 *Let (u, p) and (u_h, p_h) be the solutions of (3.4) and (3.15), respectively. Then, under the assumptions of Theorem 3.4, it holds*

$$\|u - u_h\|_0 \leq \kappa h^2(\|u\|_2 + \|p\|_1 + \|f\|_1). \tag{3.50}$$

Proof Multiplying (3.1) and (3.2) by $\Phi_h^* \in X_h^*$ and $\Psi_h \in M_h$, integrating over the dual elements V and the primary elements K, respectively, and adding the resulting equations to (3.15) with $(v_h, q_h) = (\Phi_h, \Psi_h)$, we see that

$$A(e, \Phi_h^*) + D(\Phi_h^*, \eta) + d(e, \Psi_h) + S(\eta, \Psi_h)$$
$$+ b(e, u, \Phi_h^*) + b(u, e, \Phi_h^*) - b(e, e, \Phi_h^*) = S(p, \Psi_h), \tag{3.51}$$

where $(e, \eta) = (u - u_h, p - p_h)$. Subtracting (3.51) from (3.47) with $(v, q) = (e, \eta)$ and using (3.1), we obtain

$$
\begin{aligned}
\|e\|_0^2 &= a(e, \Phi - \Phi_h) + d(e, \Psi - \Psi_h) - d(\Phi - \Phi_h, \eta) - S(\eta, \Psi_h) + S(p, \Psi_h) \\
&\quad + a(e, \Phi_h) - A(e, \Phi_h^*) - d(\Phi_h, \eta) - D(\Phi_h^*, \eta) \\
&\quad + b(u, e, \Phi - \Phi_h^*) + b(e, u, \Phi - \Phi_h^*) + b(e, e, \Phi_h^*) \\
&= a(e, \Phi - \Phi_h) + d(e, \Psi - \Psi_h) - d(\Phi - \Phi_h, \eta) - S(\eta, \Psi_h) + S(p, \Psi_h) \\
&\quad + b(u, e, \Phi - \Phi_h^*) + b(e, u, \Phi - \Phi_h^*) + b(e, e, \Phi_h^*) \\
&\quad + (f - (u \cdot \nabla)u, \Phi_h - \Phi_h^*).
\end{aligned}
\tag{3.52}
$$

Each of terms in (3.52) is estimated. Owing to the regularity (3.48) of the dual problem, we get

$$
\begin{aligned}
&|a(e, \Phi - \Phi_h) + d(e, \Psi - \Psi_h) - d(\Phi - \Phi_h, \eta)| \\
&\leq C \left(\|e\|_1 + \|\eta\|_0 \right) \left(\|\Phi - \Phi_h\|_1 + \|\Psi - \Psi_h\|_0 \right) \\
&\leq Ch^2 \left(\|u\|_2 + \|p\|_1 \right) \left(\|\Phi\|_2 + \|\Psi\|_1 \right) \\
&\leq Ch^2 \left(\|u\|_2 + \|p\|_1 \right) \|e\|_0.
\end{aligned}
$$

By using (2.17),

$$
\begin{aligned}
|S(\eta, \Psi_h) - S(p, \Psi_h)| &\leq Ch \left(\|p - \Pi_h p\|_0 + \|\eta\|_0 \right) \|\Psi\|_1 \\
&\leq Ch^2 \left(\|u\|_2 + \|p\|_1 \right) \|e\|_0.
\end{aligned}
$$

According to the bound (3.10) for the trilinear term, we have

$$
\begin{aligned}
|b(u, e, \Phi - \Phi_h^*) + b(e, u, \Phi - \Phi_h^*)| &\leq C\|u\|_2\|e\|_1 \left(\|\Phi_h - \Phi_h^*\|_0 + \|\Phi - \Phi_h\|_0 \right) \\
&\leq Ch^2 \left(\|u\|_2 + \|p\|_1 \right) \|\Phi\|_1 \\
&\leq Ch^2 \left(\|u\|_2 + \|p\|_1 \right) \|e\|_0.
\end{aligned}
$$

Similarly, applying the generalized Hölder inequality yields that

$$
\begin{aligned}
|b(e, e, \Phi_h^*)| &= |b(e, e, \Phi_h^* - \Phi_h) + b(e, e, \Phi_h)| \\
&\leq C \left(\|e\|_{L^4}\|e\|_1\|\Phi_h^* - \Phi_h\|_{L^4} + \|e\|_1^2\|\Phi_h\|_1 \right) \\
&\leq Ch^2 \left(\|u\|_2 + \|p\|_1 \right) \|e\|_0.
\end{aligned}
$$

As an immediate consequence, we derive the following important result by using a similar argument as for (2.43):

$$|(f - (u \cdot \nabla)u, \Phi_h - \Phi_h^*)|$$
$$= |([f - \hat{\pi}_h f] - [(u \cdot \nabla)u - \hat{\pi}_h(u \cdot \nabla)u], \Phi_h - \Phi_h^*)|$$
$$\leq Ch^2(\|f\|_1 + \|[(u \cdot \nabla)u]\|_1)\|\Phi_h\|_1$$
$$\leq Ch^2(\|f\|_1 + \|u\|_{L^\infty}\|u\|_2 + \|u\|_{1,4}^2)\|e\|_0$$
$$\leq Ch^2(\|f\|_1 + \|u\|_0^{1-\frac{d}{4}}\|u\|_2^{\frac{d+4}{4}} + \|u\|_1^{\frac{d}{2}}\|u\|_2^{2-\frac{d}{2}})\|e\|_0. \tag{3.53}$$

Finally, combining all these inequalities with (3.52) yields (3.50). $\square$

3.2.5 Optimal L^∞ Analysis

In this subsection, we derive an optimal estimate in the L^∞-norm for the above finite volume approximation for the stationary Navier–Stokes equations. The usual logarithmic factor $O(|\log h|)$ [13, 46, 57] is removed in this estimate by using weighted Sobolev norms. As for the stationary Navier–Stokes equations, we still need to deal with the complexity of the trilinear terms and different test and trial functions in different finite dimensional spaces.

3.2.5.1 Stability in the L^∞-Norm

Here, we collect some basic assumptions on regularity results and properties of Green's function for the Stokes equations from the literature and use them to introduce a branch of nonsingular solutions in the L^∞ norm. To present the stability of the finite volume methods in the L^∞-norm, following [13, 110], we define the regularized Green's functions for the stationary Navier–Stokes equations as follows:

$$a(G, v) - d(v, Q) + b(v, u, G) + b(u, v, G) = -(D\delta_M, v) \ \forall v \in X, \tag{3.54}$$
$$d(G, q) = 0 \ \forall q \in M. \tag{3.55}$$

Similarly, there hold the following estimates ([57], Corollary 3.4 and Theorem 3.6):

$$\|\sigma^{\mu/2-1}\nabla G\|_0 + \|\sigma^{\mu/2-1}Q\|_0 \leq Ch^{\theta/2-1}, \tag{3.56}$$
$$\|\sigma^{\mu/2}\Delta G\|_0 + \|\sigma^{\mu/2}\nabla Q\|_0 \leq Ch^{\theta/2-1}, \tag{3.57}$$

where $\sigma(x) = [|x - x_0|^2 + (\theta_1 h)^2]^{1/2}$ ($|x - x_0| < R$, $R > 0$), $\mu = 2 + \theta$, with $0 < \theta < 1$, and $C > 0$ is independent of the constant $\theta_1 > 1$ and the mesh size h.

We now consider the finite volume approximations for the stationary Navier–Stokes equations approximated by the macroelement pairs. The corresponding finite volume variational formulation is to seek $(u_h, p_h) \in X_h \times M_h$ such that

$$A(u_h, v_h^*) + D(v_h^*, p_h) + d(u_h, q_h) + b(u_h, u_h, v_h^*) = (f, v_h^*) \ \forall (v_h, q_h) \in X_h \times M_h.$$
$$(3.58)$$

The stability in terms of $\|\nabla u_h\|_{L^\infty}$ and $\|p_h\|_{L^\infty}$ is obtained by the following analysis.

Lemma 3.2 *Under the assumptions of Theorem 3.4, let $(u_h, p_h) \in X_h \times M_h$ be the solution of (3.15). Then it holds that*

$$\|\nabla u_h\|_{L^\infty} \le C(\|\nabla u\|_{L^\infty} + \|p\|_{L^\infty} + \|f\|_0). \qquad (3.59)$$

Proof Taking $(v, q) = (u_h, p_h)$ in (3.54) and (3.55), we see that

$$\|\nabla u_h\|_{L^\infty} = a(G, u_h) - d(u_h, Q) - d(G, p_h) + b(u_h, u, G) + b(u, u_h, G),$$

which, together with the Stokes projection defined in (2.57), yields

$$\|\nabla u_h\|_{L^\infty} = a(G_h, u_h) - d(u_h, Q_h) - d(G_h, p_h)$$
$$+ b(u_h, u, G) + b(u, u_h, G). \qquad (3.60)$$

Moreover, using the equivalence Lemma 2.3, it follows from the continuous and discrete variational formulations (3.4) and (3.15) approximated by the macroelement pairs that

$$a(u - u_h, v_h) - d(v_h, p - p_h) + d(u - u_h, q_h) + b(u, u, v_h) - b(u_h, u_h, v_h^*)$$
$$= (f, v_h - v_h^*). \quad (3.61)$$

Thus, we derive from (3.60) and (3.61) with $(v_h, q_h) = (G_h, -Q_h)$ that

$$\|\nabla u_h\|_{L^\infty} = a(u, G_h) - d(G_h, p) + b(u, u, G_h) - b(u_h, u_h, G_h^*)$$
$$+ b(u_h, u, G) + b(u, u_h, G) - (f, G_h - G_h^*)$$
$$= a(u, G_h - G) + a(u, G) - d(G_h - G, p) - (f, G_h - G_h^*)$$
$$+ b(u, u, G_h) - b(u_h, u_h, G_h^*) + b(u_h, u, G)$$
$$+ b(u, u_h, G), \qquad (3.62)$$

since $d(u, Q_h) = 0$. Applying (3.54) with $v = u$ and (3.2) leads to

$$a(u, G) = -2b(u, u, G) - (D\delta_M, u). \qquad (3.63)$$

Then using (3.62) and (3.63) gives the main equality

$$
\begin{aligned}
\|\nabla u_h\|_{L^\infty} = & -(D\delta_M, u) + a(u, G_h - G) - d(G_h - G, p) - (f, G_h - G_h^*) \\
& + b(u - u_h, u, G_h) - b(u - u_h, u - u_h, G_h) + b(u, u - u_h, G_h) \\
& + b(u_h, u_h, G_h - G_h^*) + b(u_h, u, G) + b(u, u_h, G) \\
& - 2b(u, u, G).
\end{aligned}
\tag{3.64}
$$

By the definition of $D\delta_M$,

$$
-(D\delta_M, u) = \|\nabla u\|_{L^\infty}.
$$

Obviously, it follows from estimate (2.24) of the functions between the finite element and finite volume spaces, the same approach as for (2.43), and the Hölder inequality that

$$
|(f, G_h - G_h^*)| \le Ch\|f\|_0(\|\nabla G\|_0 + \|Q\|_0).
$$

Moreover, the Hölder inequality yields that

$$
|a(u, G_h - G) - d(G_h - G, p)| \le (\|\nabla u\|_{L^\infty} + \|p\|_{L^\infty})\|\nabla(G - G_h)\|_{L^1}.
$$

Similarly, we estimate the trilinear terms as follows:

$$
\begin{aligned}
|b(u - u_h, u, G_h) + b(u, u - u_h, G_h)| & \le C\|\nabla u\|_0(\|\nabla G\|_0 + \|Q\|_0)\|\nabla(u - u_h)\|_0, \\
|b(u - u_h, u - u_h, G_h)| & \le C(\|\nabla G\|_0 + \|Q\|_0)\|\nabla(u - u_h)\|_0^2, \\
|b(u_h, u, G) + b(u, u_h, G) - 2b(u, u, G)| & = |b(u_h - u, u, G) + b(u, u_h - u, G)| \\
& \le C\|\nabla u\|_0\|\nabla G\|_0\|\nabla(u - u_h)\|_0.
\end{aligned}
$$

By the similar approach as for (3.39), it follows that

$$
\begin{aligned}
|b(u_h, u_h, G_h - G_h^*)| & \le Ch^{3/2}\|u_h\|_0\|A_h u_h\|_0(\|\nabla G\|_0 + \|Q\|_0) \\
& \le Ch^{3/2}\|f\|_0(\|\nabla G\|_0 + \|Q\|_0).
\end{aligned}
\tag{3.65}
$$

Using the estimate for $\|\nabla(G - G_h)\|_{L^1}$ in (2.69), together with Theorem 3.4, (3.64), and (3.65), we obtain the desired result. $\square$

It is important to note that the stability of pressure in the L^∞-norm does not directly follow from the above result on velocity and the discrete *inf-sup* condition. The analysis for pressure requires a different regularized Green's function [13, 110]:

$$
a(U, v) + d(v, V) + b(v, u, U) + b(u, v, U) = 0, \quad v \in X, \tag{3.66}
$$

$$
d(U, q) = (\delta_M - B, q), \quad q \in M, \tag{3.67}
$$

where B is a fixed function in $C_0^\infty(\Omega)$ such that $\int_\Omega B(x)dx = 1$ and thus $\delta_M - B \in L_0^2(\Omega)$. Analogically, the solution of problem (3.66) satisfies ([57], Corollary 3.4)

$$\|\sigma^{\mu/2-1}\nabla U\|_0 + \|\sigma^{\mu/2-1}V\|_0 \leq Ch^{\theta/2-1}. \tag{3.68}$$

Based on the above preparation, we need to estimate $\|p_h\|_{L^\infty}$ in Lemma 3.3.

Lemma 3.3 *Let $(u_h, p_h) \in X_h \times M_h$ be the solution of (3.15). Then it holds that*

$$\|p_h\|_{L^\infty} \leq C(\|\nabla u\|_{L^\infty} + \|p\|_{L^\infty} + \|f\|_0). \tag{3.69}$$

Proof Taking $(v, q) = (u - u_h, p_h - p)$ in (3.66) and (3.67), we find that

$$(\delta_M - B, p_h - p) = a(U, u - u_h) + d(u - u_h, V) + d(U, p_h - p)$$
$$+ b(u - u_h, u, U) + b(u, u - u_h, U). \tag{3.70}$$

Moreover, setting $(v_h, q_h) = (U_h, V_h)$ in (3.61) yields

$$a(u - u_h, U_h) - d(U_h, p - p_h) + d(u - u_h, V_h) + b(u, u, U_h)$$
$$- b(u_h, u_h, U_h^*) = (f, U_h - U_h^*). \tag{3.71}$$

Then, using (3.70) and (3.71) and noting that the Stokes projection $U_h \in X_h$ satisfies

$$d(U - U_h, p_h) = 0$$

in (2.76), we obtain

$$\|p_h\|_{L^\infty} = a(u - u_h, U - U_h) + d(u - u_h, V - V_h) - d(U - U_h, p) - (B, p - p_h)$$
$$+ (\delta_M, p) - b(u - u_h, u, U_h) - b(u_h, u - u_h, U_h) - b(u_h, u_h, U_h - U_h^*)$$
$$+ b(u - u_h, u, U) + b(u, u - u_h, U) + (f, U_h - U_h^*). \tag{3.72}$$

Using the estimate for $\|u_h\|_{L^\infty}$ in Lemma 3.2, the Hölder inequality, the estimate of $\|\nabla(U - U_h)\|_{L^1} + \|V - V_h\|_{L^1}$, and (2.77), we see that

$$|a(u - u_h, U_h - U) - d(u - u_h, V_h - V) + d(U_h - U, p)|$$
$$\leq (\|\nabla(u - u_h)\|_{L^\infty} + \|p\|_{L^\infty})(\|\nabla(U_h - U)\|_{L^1} + \|V_h - V\|_{L^1})$$
$$\leq (\|\nabla u\|_{L^\infty} + \|p\|_{L^\infty} + \|\nabla u_h\|_{L^\infty})(\|\nabla(U_h - U)\|_{L^1} + \|V_h - V\|_{L^1}).$$

Applying the estimates of the trilinear terms in (3.9) and the bound for $\|\nabla U_h\|_0$ in (2.77) yields

$$|b(u - u_h, u, U) + b(u, u - u_h, U)|$$
$$\leq C\|\nabla u\|_0 \|\nabla(u - u_h)\|_0 \|\nabla U\|_0,$$
$$|b(u - u_h, u, U_h) + b(u_h, u - u_h, U_h)|$$
$$\leq C\|\nabla(u - u_h)\|_0 (\|\nabla u\|_0 + \|u_h\|_1)(\|\nabla U\|_0 + \|V\|_0).$$

By the same approach as for (3.65),

$$|b(u_h, u_h, U_h - U_h^*)| \leq Ch^{3/2}\|f\|_0(\|\nabla U\|_0 + \|V\|_0).$$

In addition, using the Hölder inequality gives

$$|(B, p - p_h) - (\delta_M, p)| \leq C(\|u\|_{L^\infty} + \|p\|_{L^\infty}).$$

Again, we have

$$|(f, U_h - U_h^*)| \leq Ch\|f\|_0(\|\nabla U\|_0 + \|V\|_0).$$

Therefore, combining all these inequalities with an optimal estimate for $\|u - u_h\|_1$ yields the desired result. $\qquad\qquad\qquad\qquad\qquad\qquad\qquad\qquad\qquad\qquad\qquad\Box$

3.2.5.2   Optimal L^∞ Estimates

Based on the previous analysis, we will show optimal estimates in the L^∞-norm for the stationary Navier–Stokes equations.

Lemma 3.4 *Under the assumptions of Theorem 3.4, let* $(u, p) \in X \times M$ *and* $(u_h, p_h) \in X_h \times M_h$ *be the solutions of* (3.4) *and* (3.15), *respectively. Then it holds*

$$\|\nabla(u - u_h)\|_{L^\infty} \leq Ch(|u|_{2,\infty} + |p|_{1,\infty} + \|f\|_1). \qquad (3.73)$$

Proof Taking $v = e_h = I_h u - u_h$ in (3.54), recalling the properties of δ_M, and using the definition of the Stokes projection (2.57), we see that

$$\|\nabla e_h\|_{L^\infty} = a(G, e_h) - d(e_h, Q) + b(e_h, u, G) + b(u, e_h, G)$$
$$= a(G_h, e_h) - d(e_h, Q_h) + b(e_h, u, G) + b(u, e_h, G). \qquad (3.74)$$

By (2.57) and (2.55),

$$d(G_h, p - p_h) = d(G_h, p - J_h p) = d(G_h - G, p - J_h p).$$

Also, it follows from the Stokes projection (2.57) related to $(G_h, Q_h) \in X_h \times Q_h$, the difference (3.61) between the finite element methods and the finite volume methods with $(v_h, q_h) = (G_h, 0)$, and (3.55) that

$$
\begin{aligned}
a(e_h, G_h) &= a(I_h u - u, G_h) + d(G_h, p - p_h) - b(u, u, G_h) + b(u_h, u_h, G_h^*) \\
&\quad + (f, G_h - G_h^*) \\
&= a(I_h u - u, G_h - G) + a(I_h u - u, G) + d(G_h - G, p - J_h p) \\
&\quad - b(u, u, G_h) + b(u_h, u_h, G_h^*) + (f, G_h - G_h^*).
\end{aligned}
$$

Furthermore, a consequence of (3.54) with $v = I_h u - u$ is

$$
\begin{aligned}
a(I_h u - u, G) &= -(D\delta_M, I_h u - u) + d(I_h u - u, Q) - b(I_h u - u, u, G) \\
&\quad - b(u, I_h u - u, G).
\end{aligned}
$$

Thus, noting $Q_h \in M_h \subset \bar{M}_h$, using all these equations, and (3.74), we obtain

$$
\begin{aligned}
\|\nabla e\|_{L^\infty} &= a(I_h u - u, G_h - G) + d(G_h - G, p - J_h p) + d(I_h u - u, Q - Q_h) \\
&\quad - d(e, Q_h) - (D\delta_M, I_h u - u) + (f, G_h - G_h^*) \\
&\quad - b(I_h u - u, u, G) - b(u, I_h u - u, G) - b(u, u, G_h) + b(u_h, u_h, G_h) \\
&\quad - b(u_h, u_h, G_h) + b(u_h, u_h, G_h^*) + b(e, u, G) + b(u, e, G). \quad (3.75)
\end{aligned}
$$

Clearly, it follows from the definition of δ_M that

$$
-(D\delta_M, I_h u - u) = \|\nabla(I_h u - u)\|_{L^\infty}.
$$

The Hölder inequality gives

$$
\begin{aligned}
&|a(I_h u - u, G_h - G) + d(G_h - G, p - J_h p) + d(I_h u - u, Q - Q_h)| \\
&\leq (\|\nabla(I_h u - u)\|_{L^\infty} + \|p - J_h p\|_{L^\infty})(\|\nabla(G_h - G)\|_{L^1} + \|Q - Q_h\|_{L^1}), \\
&|(f, G_h - G_h^*)| = |(f - \hat{\pi}_h f, G_h - G_h^*)| \leq Ch^2 \|f\|_1 \|\nabla G\|_0.
\end{aligned}
$$

Using the estimates of the trilinear terms yields that

$$
\begin{aligned}
|b(I_h u - u, u, G) + b(u, I_h u - u, G)| &\leq C\|I_h u - u\|_0 \|Au\|_0 \|\nabla G\|_0, \\
|b(e, u, G) + b(u, e, G)| &\leq C(\|u - u_h\|_0 + \|I_h u - u\|_0)\|Au\|_0 \|\nabla G\|_0.
\end{aligned}
$$

Furthermore, using the property of the projection operator (2.57) gives

$$
\begin{aligned}
&|b(u, u, G_h) - b(u_h, u_h, G_h)| \\
&= |b(u - u_h, u, G_h) + b(u, u - u_h, G_h) - b(u - u_h, u - u_h, G_h)| \\
&\leq C(\|Au\|_0 \|u - u_h\|_0 + \|\nabla(u - u_h)\|_0^2)(\|\nabla G\|_0 + \|Q\|_0). \quad (3.76)
\end{aligned}
$$

In view of Lemma 2.2 and the Hölder inequality, it follows that

$$
\begin{aligned}
&|b(u_h, u_h, G_h - G_h^*)| \\
&= \left(((u_h - \hat{\pi}_h u_h) \cdot \nabla) u_h + \frac{1}{2} \mathrm{div}\, u_h (u_h - \hat{\pi}_h u_h), G_h - G_h^* \right) \\
&\leq \left(1 + \frac{\sqrt{d}}{2} \right) \|\nabla u_h\|_{L^\infty} \|u_h - \hat{\pi}_h u_h\|_0 \|G_h - G_h^*\|_0 \\
&\leq Ch^2 \|\nabla u_h\|_{L^\infty} (\|\nabla G\|_0 + \|Q\|_0).
\end{aligned}
\tag{3.77}
$$

Using the estimates of $\|\nabla G\|_0$, $\|Q\|_0$, $\|\nabla u_h\|_{L^\infty}$, $\|\nabla(G - G_h)\|_{L^1}$ and $\|Q - Q_h\|_{L^1}$ again, we find that

$$
\|\nabla(u - u_h)\|_{L^\infty} \leq \|\nabla(I_h u - u)\|_{L^\infty} + \|\nabla e\|_{L^\infty},
$$

which, together with (3.75)–(3.77), Theorem 3.4, Lemma 3.2, and the following interpolation property ([37], Theorem 3.1.6)

$$
\|\nabla(I_h u - u)\|_{L^\infty} + \|p - J_h p)\|_{L^\infty} \leq Ch \|u\|_{2,\infty},
\tag{3.78}
$$

gives the desired result. $\square$

It is worth noticing that the analysis for $\|p - p_h\|_{L^\infty}$ still requires the stability result in the L^∞-norm for velocity and pressure, the Stokes projection (U_h, V_h), and assumption (**A3**). Then, the proof of the best approximation property will be given in the next theorem.

Lemma 3.5 *Let* $(u, p) \in X \times M$ *and* $(u_h, p_h) \in X_h \times M_h$ *be the solutions of* (3.4) *and* (3.15), *respectively. Then it holds*

$$
\|p - p_h\|_{L^\infty} \leq Ch(|u|_{2,\infty} + |p|_{1,\infty} + \|f\|_1).
\tag{3.79}
$$

Proof Using (3.70) and (3.71) and setting $\eta = J_h p - p_h$ give

$$
\begin{aligned}
\|\eta\|_{L^\infty} =\ & a(u - u_h, U - U_h) + d(u - u_h, V - V_h) - d(U - U_h, p - J_h p) \\
& + (B, p - p_h) + (\delta_M, J_h p - p) - b(u - u_h, u, U_h) + b(u - u_h, u - u_h, U_h) \\
& - b(u, u - u_h, U_h) - b(u_h, u_h, U_h - U_h^*) + b(u - u_h, u, U) \\
& + b(u, u - u_h, U) + (f, U_h - U_h^*).
\end{aligned}
\tag{3.80}
$$

Owing to Theorem 3.2, Lemma 3.4, the Hölder inequality, (3.78) and using the bound of $\|\nabla(U - U_h)\|_{L^1}$ and $\|V - V_h\|_{L^1}$ again, we see that

$$
\begin{aligned}
|a(u &- u_h, U_h - U) - d(u - u_h, V_h - V) + d(U_h - U, p - J_h p)| \\
&\leq (\|\nabla(u - u_h)\|_{L^\infty} + \|p - J_h p\|_{L^\infty})(\|\nabla(U_h - U)\|_{L^1} + \|V_h - V\|_{L^1}) \\
&\leq Ch(|u|_{2,\infty} + |p|_{1,\infty} + \|f\|_1)(\|\nabla(U_h - U)\|_{L^1} + \|V_h - V\|_{L^1}).
\end{aligned}
$$

By (2.10), Theorem 3.4, and a simple calculation, we obtain

$$
\begin{aligned}
|(B, p - p_h) + (\delta_M, J_h p - p)| &\leq C(\|p - p_h\|_0 + \|p - J_h p\|_{L^\infty}) \\
&\leq Ch(\|u\|_{2,\infty} + |p|_{1,\infty} + \|f\|_1).
\end{aligned}
$$

As for the trilinear terms, we deduce from (3.8)–(3.10) that

$$
\begin{aligned}
|b(u - u_h, u, U) + b(u, u - u_h, U)| &\leq C\|Au\|_0 \|u - u_h\|_0 \|\nabla U\|_0, \\
|b(u - u_h, u, U_h) + b(u, u - u_h, U_h)| &\leq C\|Au\|_0 \|u - u_h\|_0 (\|\nabla U\|_0 + \|V\|_0), \\
|b(u - u_h, u - u_h, U_h)| &\leq C\|\nabla(u - u_h)\|_0^2 (\|\nabla U\|_0 + \|V\|_0).
\end{aligned}
$$

Applying the estimates in (3.77) gives

$$
|b(u_h, u_h, U_h - U_h^*)| \leq Ch^2 \|\nabla u_h\|_{L^\infty}(\|\nabla U\|_0 + \|V\|_0).
$$

In order to compensate for the order $O(h^{-1})$ of $\|\nabla U\|_0$ and $\|V\|_0$, we improve the regularity of the right-hand side $f \in [H^1(\Omega)]^d$ and combine

$$
\begin{aligned}
|b(u_h, u_h, U_h - U_h^*)| &\leq Ch^2 \|\nabla u_h\|_{L^\infty}(\|\nabla U\|_0 + \|V\|_0), \\
|(f, U_h - U_h^*)| &\leq Ch^2 \|f\|_1 (\|\nabla U\|_0 + \|V\|_0).
\end{aligned}
$$

Finally, combining all these inequalities with the estimates of $\|\nabla(U - U_h)\|_{L^1}$, $\|V - V_h\|_{L^1}$, $\|\nabla U\|_0$ and $\|V\|_0$ and using (2.53), Theorem 3.4 and Lemma 3.2, we obtain the desired estimate. $\qquad\square$

Now, the main result in the L^∞-norm for velocity and pressure is summarized in the next theorem.

Theorem 3.6 *Let $(u, p) \in X \times M$ and $(u_h, p_h) \in X_h \times M_h$ be the solutions of (3.4) and (3.15), respectively. Then it holds*

$$
\|\nabla(u - u_h)\|_{L^\infty} + \|p - p_h\|_{L^\infty} \leq Ch(|u|_{2,\infty} + |p|_{1,\infty} + \|f\|_1).
$$

3.3 FVEs of Branches of Nonsingular Solutions

In this section, we concentrate on practical examples whose solutions are mostly isolated; i.e., there exists a neighborhood in which each solution is unique. As the viscosity varies along an interval, each solution of the Navier–Stokes equations describes an isolated branch since the solutions depend continuously on the viscosity. In particular, this means that the bifurcation phenomenon is rare. This situation, very frequently encountered in practice, is expressed mathematically by the notion of branches of nonsingular solutions. Several papers proposed and analyzed the finite element approximations of branches of nonsingular solutions pertaining to the stationary Navier–Stokes equations based on a general form of an implicit function theorem, which is a variant of a broader theory developed by Brezzi et al. [16, 59]. This book introduces the corresponding finite volume approximations of branches of nonsingular solutions [86–89].

3.3.1 An Abstract Framework

A linear operator $T : Y \to \bar{X}$ is defined as follows: Given $g \in Y$, the solution of the Stokes problem

$$\begin{aligned} -\Delta v + \lambda \nabla q &= g, && \text{in } \Omega, \\ \operatorname{div} v &= 0, && \text{in } \Omega, \\ v &= 0, && \text{on } \partial\Omega \end{aligned}$$

is denoted by $\tilde{v}(\lambda) = (v, \lambda q) = Tg \in \bar{X}$. Furthermore, a C^2-mapping $\mathscr{G} : R^+ \times \bar{X} \to Y$ is defined by

$$\mathscr{G}(\lambda, \tilde{v}(\lambda)) = \lambda \left((v \cdot \nabla)v + \frac{1}{2}(\operatorname{div} v)v - f \right).$$

Finally, we define

$$F(\lambda, \tilde{v}(\lambda)) = \tilde{v}(\lambda) + T\mathscr{G}(\lambda, \tilde{v}(\lambda)), \quad \lambda \in R^+, \ \tilde{v}(\lambda) \in \bar{X}.$$

In this section, a branch of nonsingular solutions of the stationary Navier–Stokes equations, as introduced in [16, 59], are studied. Let Λ be a compact interval in R^+; $\{(\lambda, \tilde{u}(\lambda))\}$, with $\tilde{u}(\lambda) = (u, \lambda p)$, is a branch of nonsingular solutions to the equation

$$F(\lambda, \tilde{u}(\lambda)) = 0, \tag{3.81}$$

if $D_u F(\lambda, \tilde{u}(\lambda))$ is an isomorphism from $\bar{X}$ onto Y for all $\lambda \in \Lambda$.

Using integration by parts, the weak formulation of the stationary Navier–Stokes equations (3.1)–(3.3) is: Find $(u, p) \in \bar{X}$ such that

$$a(u, v) - \lambda d(v, p) + \lambda d(u, q) + \lambda b(u, u, v) = \lambda(f, v) \quad \forall (v, q) \in \bar{X}. \quad (3.82)$$

Furthermore, the existence and uniqueness results of (3.82) can be obtained [59, 67, 86–88].

Lemma 3.6 (Theorem 2.1 in [88] and Lemma 2.1 in [86]) *If λ satisfies the following uniqueness condition:*

$$\lambda < \lambda_0 = \frac{1}{\sqrt{C_{10} \|f\|_{-1}}},$$

then (3.82) *admits a unique solution* (u, p). *Moreover, the pair* $(u, p) \in \bar{X}$ *is a solution of the problem* (3.82) *if and only if* $u(\lambda) \in \bar{X}$ *is a solution of* (3.81).

Similarly, we can apply the same approach as in [67] to obtain the following stability of (3.82):

Lemma 3.7 *Assume that $\partial \Omega$ has a sufficient smooth boundary in C^2, $f \in Y$, and the pair $\tilde{u}(\lambda) = (u, \lambda p) \in \bar{X}$ is a solution of problem* (3.81). *Then $u(\lambda) \in D(A) \times [H^1(\Omega) \cap M]$ and $\mathscr{G}(\lambda, u(\lambda)) \in Y$ satisfy*

$$\|u\|_2 + \lambda \|p\|_1 \leq C. \quad (3.83)$$

Accordingly, set $\bar{X}_h \equiv X_h \times M_h$. Then, a bilinear form on $\bar{X}_h \times \bar{X}_h$ for the finite element method introduced in [118, 120] is rewritten by

$$\mathscr{B}_\lambda((\bar{u}_h, \lambda \bar{p}_h), (v_h, \lambda q_h)) = a(\bar{u}_h, v_h) - \lambda d(v_h, \bar{p}_h) + \lambda d(\bar{u}_h, q_h) + \lambda G(p_h, q_h) \quad (3.84)$$
$$\forall (\bar{u}_h, \bar{p}_h), (v_h, q_h) \in \bar{X}_h.$$

This bilinear form satisfies the continuity and weak coercivity properties [122]:

$$|\mathscr{B}_\lambda((\bar{u}_h, \lambda \bar{p}_h), (v_h, \lambda q_h))| \leq C \|(\bar{u}_h, \lambda \bar{p}_h)\| \| (v_h, \lambda q_h)\|, \quad (3.85)$$
$$\sup_{(v_h, \lambda q_h) \in \bar{X}_h} \frac{|\mathscr{B}_\lambda((\bar{u}_h, \lambda \bar{p}_h), (v_h, \lambda q_h))|}{\|(v_h, \lambda q_h)\|} \geq \beta_\lambda \|(\bar{u}_h, \lambda \bar{p}_h)\|, \quad (3.86)$$

where the constant $\beta_\lambda > 0$ is independent of h.

Using the above notation, the corresponding finite element formulation of system (3.81) reads: Find $(\bar{u}_h, \bar{p}_h) \in \bar{X}_h$, such that, for all $(v_h, q_h) \in \bar{X}_h$,

$$\mathscr{B}_\lambda((\bar{u}_h, \lambda\bar{p}_h), (v_h, \lambda q_h)) + \lambda b(\bar{u}_h, \bar{u}_h, v_h) = \lambda(f, v_h); \tag{3.87}$$

i.e.,

$$F(\lambda, \tilde{\bar{u}}_h(\lambda)) \equiv \tilde{\bar{u}}_h(\lambda) + T_h\mathscr{G}(\lambda, \tilde{\bar{u}}_h(\lambda)) = 0, \tag{3.88}$$

where T_h is the discrete counterpart of the operator T.

Now, the finite volume variational formulation for the stationary Navier–Stokes equations (3.81) is: Find $\tilde{u}_h(\lambda) = (u_h, \lambda p_h) \in \bar{X}_h \subset \bar{X}$ such that

$$F_h(\lambda, \tilde{u}_h(\lambda)) \equiv \tilde{u}_h(\lambda) + T_h\mathscr{G}(\lambda, \tilde{u}_h(\lambda)) = 0; \tag{3.89}$$

i.e.,

$$\mathscr{C}_\lambda((u_h, \lambda p_h), (v_h, \lambda q_h)) + \lambda b(u_h, u_h, v_h^*) = \lambda(f, v_h^*), \forall(v_h, q_h) \in \bar{X}_h,$$

where the bilinear form $\mathscr{C}_h(\cdot, \cdot)$ on $\bar{X}_h \times \bar{X}_h$ is

$$\mathscr{C}_\lambda((u_h, \lambda p_h), (v_h, \lambda q_h)) = A(u_h, v_h^*) + \lambda D(v_h^*, p_h) + \lambda d(u_h, q_h)$$
$$+ \lambda G(p_h, q_h). \tag{3.90}$$

Then, the continuity and weak coercivity of the new bilinear form $\mathscr{C}_\lambda(\cdot, \cdot)$ can be easily verified [88]:

$$|\mathscr{C}_\lambda((u_h, \lambda p_h), (v_h, \lambda q_h))| \leq C |\!|\!|(u_h, \lambda p_h)|\!|\!|\,|\!|\!|(v_h, \lambda q_h)|\!|\!|$$
$$\forall(u_h, p_h), \ (v_h, q_h) \in \bar{X}_h, \tag{3.91}$$

and

$$\sup_{(v_h, q_h) \in \bar{X}_h} \frac{|\mathscr{C}_\lambda((u_h, \lambda p_h), (v_h, \lambda q_h))|}{|\!|\!|(v_h, \lambda q_h)|\!|\!|} \geq \beta_\lambda^* |\!|\!|(u_h, \lambda p_h)|\!|\!| \ \forall(u_h, p_h) \in \bar{X}_h, \tag{3.92}$$

where the constant $\beta_\lambda^* > 0$ is independent of h.

3.3.2 Existence and Uniqueness

In this subsection, the main goal is to provide the existence and uniqueness of a branch of nonsingular solutions of the finite volume methods for the stationary Navier–Stokes equations. Due to the complexity of the nonlinear Navier–Stokes problem, the Brouwer fixed theory is applied in establishing the stability of the finite volume solution for this problem. A similar proof can be performed as in Theorem 3.2.

Lemma 3.8 *Assume that Lemma 3.7 holds. Then problem* (3.89) *has a set of solutions* $\tilde{u}_h(\lambda) = (u_h, \lambda p_h) \in \bar{X}_h$ *satisfying* (3.18).

3.3.3 Optimal Analysis

In this subsection, an optimal analysis related to a branch of nonsingular solutions of the finite volume methods is provided for the stationary Navier–Stokes equations with large data.

3.3.3.1 Optimal H^1 and L^∞ Estimates

Similar to the continuous case, $\{(\lambda, \tilde{\bar{u}}_h(\lambda))\}$ with $\tilde{\bar{u}}_h(\lambda) = (\bar{u}_h, \lambda \bar{p}_h)$ is a branch of nonsingular solutions to (3.88) if

$$D_{\bar{u}_h} F(\lambda, \tilde{\bar{u}}_h(\lambda)) \text{ is an isomorphism from } \bar{X}_h \text{ onto } Y \text{ for all } \lambda \in \Lambda. \quad (3.93)$$

Recall that $T_h : Y \to \bar{X}_h$ is the solution operator of the discrete Stokes equations. This operator yields the solution $\tilde{\bar{u}}_h(\lambda) = (\bar{u}_h, \lambda \bar{p}_h)$ to problem (3.88). Apparently, this solution is also a solution of the discrete Navier–Stokes equation (3.88) if and only if it is a solution of (3.87). Furthermore, by (3.93) and the results in [16, 59, 67], we have the following proposition:

Proposition 3.1 $\tilde{\bar{u}}_h(\lambda) \in \bar{X}_h$ *is a branch of nonsingular solutions to Eq.* (3.88) *if there exists a constant* $\gamma > 0$, *dependent of the data* (λ, f, Ω), *such that*

$$\sup_{(v_h, q_h) \in \bar{X}_h} \frac{\bar{B}_\lambda((\bar{w}_h, \lambda \bar{\chi}_h); (v_h, \lambda q_h))}{\|(v_h, \lambda q_h)\|} \geq \gamma_\lambda \|(\bar{w}_h, \lambda \bar{\chi}_h)\|, (\bar{w}_h, \bar{\chi}_h) \in \bar{X}_h,$$

$$(3.94)$$

where

$$\bar{B}_\lambda((\bar{w}_h, \lambda \bar{\chi}_h); (v_h, \lambda q_h)) \equiv \bar{A}_\lambda(\bar{u}_h; \bar{w}_h, v_h) - \lambda d(v_h, \bar{\chi}_h) + \lambda d(\bar{w}_h, q_h),$$

and $\bar{A}_\lambda(\bar{u}_h; w_h, v_h) = a(w_h, v_h) + \lambda b(\bar{u}_h, w_h, v_h) + \lambda b(w_h, \bar{u}_h, v_h).$

Suppose that problem (3.81) has a branch of nonsingular solutions $\{(\lambda, \tilde{u}(\lambda)); \lambda \in \Lambda\}$ and that the following assumption (**A2**) holds:

Assumption (A2) There exists another Banach space Z contained in Y, with continuous imbedding, such that

$$D_u \mathscr{G}(\lambda, u(\lambda)) \in \pounds(\bar{X}, Z) \quad \forall \lambda \in \Lambda, \quad u \in \bar{X},$$
$$\lim_{h \to 0} \|(T_h - T)g\|_{\bar{X}} = 0 \quad \forall g \in Y,$$
$$\lim_{h \to 0} \|(T_h - T)\|_{\pounds(Z, \bar{X})} = 0.$$

Then the next result holds for the stabilized finite element methods ([67], Theorem 4.3) of the stationary Navier–Stokes equations in two dimensions, which can be easily extended to the stabilized finite element methods approximated by the lower order finite element pairs.

Theorem 3.7 *Assume that $\mathscr{G}$ is a C^2-mapping from $\Lambda \times \bar{X}$ onto Y, the mapping $D_{uu}\mathscr{G}(\lambda, \tilde{u}(\lambda))$ is bounded on all bounded subsets of $\Lambda \times \bar{X}$, the assumptions (**A1**) and (**A2**) hold, and $\{(\lambda, \tilde{u}(\lambda)); \lambda \in \Lambda\}$ is a branch of nonsingular solutions of (3.81). Then there exist a neighborhood ϑ of the origin in $\bar{X}$ and, for $0 < h \leq h_0$ small enough, a unique C^2-function $\lambda \in \Lambda \to \tilde{\bar{u}}_h(\lambda) \in \bar{X}_h$ such that*

$$\{(\lambda, \tilde{\bar{u}}_h(\lambda)); \lambda \in \Lambda\} \text{ is a branch of nonsingular solutions to (3.88),}$$
$$\tilde{\bar{u}}_h(\lambda) - \tilde{u}(\lambda) \in \vartheta \text{ for all } \lambda \in \Lambda. \quad (3.95)$$

Furthermore, there exists a constant $\kappa > 0$, independent of h, such that, for all $\lambda \in \Lambda$,

$$\|\bar{u}_h - u\|_0 + h\|\tilde{\bar{u}}_h(\lambda) - \tilde{u}(\lambda)\| \leq \kappa h(\|u\|_2 + \|p\|_1 + \|f\|_0). \quad (3.96)$$

Obviously, it shows that there is a branch of nonsingular solutions $\{(\lambda, \tilde{\bar{u}}_h(\lambda)); \lambda \in \Lambda\}$ in the neighborhood ϑ for a sufficiently small mesh scale $h > 0$ and all $\lambda \in \Lambda$. Assume that $\{(\lambda, \tilde{u}_h(\lambda)); \lambda \in \Lambda\}$ is a branch of nonsingular solutions of the finite volume methods for the stationary Navier–Stokes equations (3.89). From the geometric point of view, we now need to show that these solutions are also located in the same neighborhood ϑ.

In a similar manner as for the proof of Proposition 3.1, we give the following proposition:

Proposition 3.2 $\{(\lambda, \tilde{u}_h)\} \in \bar{X}_h$ *is a nonsingular solution to Eq.* (3.89) *if there exists constants* $\gamma^* > 0$, *dependent of the data* (λ, Ω, f), *such that*

$$\sup_{(v_h, q_h) \in \bar{X}_h} \frac{B_\lambda((w_h, \lambda \chi_h); (v_h, \lambda q_h))}{\|(v_h, \lambda q_h)\|} \geq \gamma_\lambda^* \|(w_h, \lambda \chi_h)\|, \tag{3.97}$$

where

$$B_\lambda((w_h, \lambda \chi_h); (v_h, \lambda q_h)) = A_\lambda(u_h; w_h, v_h^*) + \lambda D(v_h^*, \chi_h) + \lambda d(w_h, q_h),$$
$$A_\lambda(u_h; w_h, v_h^*) = A(w_h, v_h^*) + \lambda b(u_h, w_h, v_h^*) + \lambda b(w_h, u_h, v_h^*),$$

and γ_λ^* *is different from* γ_λ *in* (3.94).

After having briefly recalled some basic facts about classical finite element methods, we shall present the stability and convergence results for the corresponding finite volume variational formulation (3.89). The study of these problems will be the kernel of this subsection.

Theorem 3.8 *Under the assumptions of Theorem 3.7, then there exist a neighborhood* ϑ *of the origin in* $\bar{X}$ *and, for* $h \leq h_0$ *small enough, a unique* C^2-*function* $\lambda \in \Lambda \rightarrow \tilde{u}_h(\lambda) \in \bar{X}_h$ *such that*

$$\{(\lambda, \tilde{u}_h(\lambda)); \lambda \in \Lambda\} \text{ is a branch of nonsingular solutions to (3.89)},$$
$$\tilde{u}_h(\lambda) - \tilde{u}(\lambda) \in \vartheta \text{ for all } \lambda \in \Lambda. \tag{3.98}$$

Furthermore, there exists a constant $\kappa > 0$, *independent of* h *such that, for all* $\lambda \in \Lambda$,

$$\|\tilde{\tilde{u}}_h(\lambda) - \tilde{u}_h(\lambda)\| \leq \kappa h^{3/2} \|f\|_1. \tag{3.99}$$

Moreover, it holds that

$$\|\tilde{u}_h(\lambda) - \tilde{u}(\lambda)\| \leq \kappa h(\|u\|_2 + \|p\|_1 + \|f\|_0), \tag{3.100}$$
$$\|u - u_h\|_0 \leq \kappa h^2(\|u\|_2 + \|p\|_1 + \|f\|_1). \tag{3.101}$$

Proof First, applying Theorem 3.1, the reader can easily check (3.99); i.e., the finite volume solutions and the corresponding finite element solutions satisfy the superclose result (3.99). By Proposition 3.1 and the equivalence Lemma 2.3, we have the following relationship between two terms: $A_\lambda(u_h, w_h, v_h^*)$ and $A_\lambda(\bar{u}_h, w_h, v_h)$

$$A_\lambda(u_h, w_h, v_h^*) = \bar{A}_\lambda(\bar{u}_h, w_h, v_h) - \lambda b(e, w_h, v_h) - \lambda b(w_h, e, v_h)$$
$$-\lambda b(u_h, w_h, v_h - v_h^*) - \lambda b(w_h, u_h, v_h - v_h^*). \quad (3.102)$$

Then, we estimate the above equality by (3.9) and (3.33) as follows:

$$|\lambda b(e, w_h, v_h) + \lambda b(w_h, e, v_h)| \leq C\|e\|_1 \|w_h\|_1 \|v_h\|_1$$
$$\leq C\lambda h^{3/2} \|f\|_1 \|w_h\|_1 \|v_h\|_1$$
$$\leq C\lambda h^{3/2} \|f\|_1 \|(w_h, \chi_h)\| \|v_h\|_1.$$

Similarly, using the Hölder inequality and (3.18) to obtain that

$$|\lambda b(u_h, w_h, v_h - v_h^*) + \lambda b(w_h, u_h, v_h - v_h^*)|$$
$$\leq 2\lambda(\|u_h\|_{L^\infty}\|w_h\|_1 + \|w_h\|_{L^\infty}\|u_h\|_1)\|v_h - v_h^*\|_0$$
$$\leq C\lambda h^{1/2} \|u_h\|_1 \|w_h\|_1 \|v_h\|_1$$
$$\leq C\lambda^2 \gamma h^{1/2} \|f\|_0^2 \|w_h\|_1 \|v_h\|_1$$
$$\leq C\lambda^2 \gamma h^{1/2} \|f\|_0^2 \|(w_h, \chi_h)\| \|v_h\|_1.$$

Thus, choosing $\gamma^* = \gamma - 2C\lambda^2 \gamma h^{1/2}\|f\|_0^2$, substitute (3.33) and (3.102) into (3.97) to obtain, for sufficient small $h > 0$,

$$\sup_{(v_h, q_h) \in \bar{X}_h} \frac{B_\lambda((w_h, \lambda\chi_h); (v_h, \lambda q_h))}{\|(v_h, q_h)\|} \geq (\gamma - 2C\lambda^2 \gamma\|f\|_0 h^{1/2})\|(w_h, \chi_h)\|$$
$$= \gamma^* \|(w_h, \chi_h)\|. \quad (3.103)$$

Thus, we complete the proof by Proposition 3.2.

Apparently, a superconvergence result is obtained between the finite element solutions and the corresponding finite volume solutions. Using a result in [59] and the estimate between them, (3.98) still holds with respect to the solutions of the finite volume methods around the same neighborhood ϑ of the origin in $\bar{X}$. Furthermore, we now give an optimal analysis of a branch of the finite volume solutions for the stationary Navier–Stokes equations with large data.

By a triangle inequality and (3.96), we have

$$\|\tilde{u}_h(\lambda) - \tilde{u}(\lambda)\| \leq \|\tilde{u}_h(\lambda) - \bar{\tilde{u}}_h(\lambda)\| + \|\bar{\tilde{u}}_h(\lambda) - \tilde{u}(\lambda)\|$$
$$\leq \kappa h(\|u\|_2 + \|p\|_1 + \|f\|_0), \quad (3.104)$$

which completes the proof of (3.100). Also, using the Aubin–Nitsche duality technique and the same approach as for Theorem 3.5 yields the desired result (3.101).

$$\square$$

3.3.4 Optimal L^∞ Estimate

This subsection is devoted to an optimal L^∞ estimate for a branch of finite volume approximations for the full Navier–Stokes problems. The extension is entirely standard but the approximation is presented in a fairly nonlinear system. We can apply the same approach as for the results in Section 3 and some techniques of the trilinear term to obtain an optimal analysis in the L^∞-norm for velocity gradient and pressure. Here, optimal L^∞ estimates are directly given without a proof.

Theorem 3.9 *Under the assumption of Theorem 3.8, let* $\{\lambda, \tilde{u}(\lambda); \ \lambda \in \Lambda\}$ *and* $\{\lambda, \tilde{u}_h(\lambda); \ \lambda \in \Lambda\}$ *be a branch of nonsingular solutions of* (3.81) *and* (3.89), *respectively. Then it holds that*

$$\|\tilde{u}(\lambda) - \tilde{u}_h(\lambda)\|_{L^\infty} \le Ch(|u|_{2,\infty} + |p|_{1,\infty} + \|f\|_1).$$

3.4 Numerical Experiments

In this section, we present numerical experiments to check the numerical theory developed in this chapter. In all the experiments, Ω is a unit square in $\Re^2$, the viscosity $\mu = 1$, the exact solution for velocity $u = (u_1, u_2)$ and pressure p is given as follows:

$$p(x) = 10(2x_1 - 1)(2x_2 - 1),$$
$$u_1(x) = 10x_1^2(x_1 - 1)^2 x_2(x_2 - 1)(2x_2 - 1),$$
$$u_2(x) = -10x_1(x_1 - 1)(2x_1 - 1)x_2^2(x_2 - 1)^2,$$

and the right-hand side $f(x)$ is determined by (3.1).

We proceed by studying two different cases. The first one concerns the $P_1 - P_0$ case, while in the second one, we consider $P_1 - P_1$. The convergence rate of the stabilized finite volume methods is investigated for the stationary Navier–Stokes equations. The standard finite volume methods with the MINI element, the classical nonconforming finite element $NCP_1 - P_1$ and the stabilized finite volume methods presented in this section are compared for the stationary Navier–Stokes equations on uniform meshes. As seen from Tables 3.1, 3.2 and Fig. 3.1, numerical results completely support the theoretical analysis for the stationary Navier–Stokes equations.

Moreover, the errors between the stabilized finite element solutions and the finite volume solutions obtained are studied. Obviously, superclose results are obtained for velocity and pressure in the H^1-norm and L^2-norm, respectively, which agrees with Theorems 3.1 and 3.8.

Table 3.1 Superconvergence between FVM and FEM: $P_1 - P_0$

1/h	$\frac{\|u-u_h\|_0}{\|u\|_0}$	ECR	$\frac{\|u-u_h\|_1}{\|u\|_1}$	ECR	$\frac{\|p-p_h\|_0}{\|p\|_0}$	ECR
9	2.7081E-2		2.6709E-2		2.0226E-3	
18	7.29665E-3	1.8920	7.3752E-3	1.8566	5.0015E-4	2.0157
27	3.2697E-3	1.9797	3.3180E-3	1.9700	2.2300E-4	1.9921
36	1.8436E-3	1.9917	1.8729E-3	1.9880	1.2566E-4	1.9940

Table 3.2 Superconvergence between FVM and FEM: $P_1 - P_1$

1/h	$\frac{\|u-u_h\|_0}{\|u\|_0}$	ECR	$\frac{\|u-u_h\|_1}{\|u\|_1}$	ECR	$\frac{\|p-p_h\|_0}{\|p\|_0}$	ECR
1/9	2.6619E-2		2.2846E-2		2.3567E-4	
1/18	7.3113E-3	1.8643	7.2580E-3	1.6543	5.26E-05	2.1638
1/27	3.2737E-3	1.9817	3.3047E-3	1.9404	2.30E-05	2.0435
1/36	1.8450E-3	1.9933	1.8704E-3	1.9786	1.29E-05	2.0177

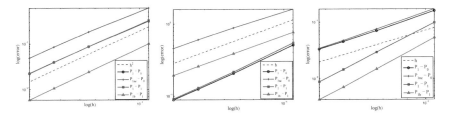

Fig. 3.1 Comparison of ECR: $P_1 - P_i$, $i = 0, 1$, $P_1b - P_1$ and $NCP_1 - P_1$

3.5   Conclusions

In this chapter, we consider the stabilized finite volume methods for the stationary Navier–Stokes equations. These methods are introduced through their relationship with the conforming finite element methods using the lower order finite element pairs. While the finite volume methods are studied for the Stokes equations [84, 117], the main difficulty we face is overcoming the difficulties of the nonlinear discrete terms arising from the finite volume discretization of the Navier–Stokes equations.

On one hand, stability and convergence analyses for the finite volume methods for the Navier–Stokes equations are presented under the uniqueness condition of the solution, which requires that the data is small enough in certain norms [95, 122]. On the other hand, the analyses depend on an abstract theory of Brezzi et al. [16], Girault and Raviart [59], and He and Li [67] for a branch of nonsingular solutions for these equations, which overcomes the uniqueness condition with small data. We perform the stability and convergence analyses for the finite volume methods for the stationary Navier–Stokes equations without relying on the unique solution condition. Optimal estimates are also obtained for velocity in the L^2- and H^1-norm and for pressure in the L^2-norm.

Although there are some results of the finite volume methods for the Stokes equations [31, 32, 53, 84, 130, 134], an analysis for these methods for the Navier–Stokes equations is lacking. In particular, there is a difficulty in handling the nonlinear discrete terms of the Navier–Stokes equations because these terms lack skew-symmetry in the context of a Petrov–Galerkin method which uses different trial and test functions in different finite dimensional spaces. Hence, an analysis for these equations must take special care of the nonlinear discrete terms arising from the finite volume discretization. Moreover, the regularity of the source term may affect a convergence rate of the finite volume methods. An optimal L^2-norm estimate for velocity is also one of the major difficulties in the analysis of the finite volume methods for these equations without any additional regularity on the original solution. Here, a new duality argument is first introduced to establish convergence of optimal order in this norm for velocity [95]. Moreover, the convergence analysis also shows an important superconvergence result between the conforming mixed finite element solution and the finite volume solution using the same finite element pairs for these equations.

Furthermore, the derivation of error estimates in the L^∞-norm is another difficult task for the analysis of the finite volume methods (even the finite element methods) for the stationary Navier–Stokes equations. These estimates bear a logarithmic factor $O(|\log h|)$ [37], where h is a grid size. The technique in this chapter in removing this factor relies on new weighted L^2-norm estimates for regularized Green's functions for the finite element methods [57] and the relationship between the finite element methods and the finite volume methods for the Stokes problem [84, 95, 96, 134]. A stability and optimal analysis in the L^∞-norm is carried out for velocity gradient and pressure for the stationary Navier–Stokes equations.

Chapter 4
FVMs for the Nonstationary Navier–Stokes Equations

Abstract In this chapter, we mainly discuss the stabilized lower order finite volume methods for the nonstationary Navier–Stokes equations. Under the same regularity assumption on the exact solution for the corresponding stabilized finite element methods, the stabilized finite volume methods provide approximate solutions with the same optimal convergence rates for u_h in the H^1 norm and p_h in the L^2 norm. Furthermore, under a slightly different regularity assumption on the source term, the stabilized finite volume methods can also reach the optimal convergence rate for u_h in the L^2 norm.

4.1 Introduction

The finite volume methods presented here are particularly designed to inherit a local conservation property associated with the differential equations. There exist two major difficulties in the convergence analysis of these stabilized finite volume methods. While the analysis can be carried out through their relationship with the conforming finite elements of the lower order pairs for the Stokes equations [84], there still exists an additional difficulty in the treatment of the nonlinear term (i.e., the trilinear term) appearing in a finite volume formulation of the Navier–Stokes equations. The other major difficulty is associated with the analysis of the discretization of the transient term. Because an equivalent operator between the finite volume methods and the $P_1 - P_i$, $i = 1, 0$, pairs of the conforming finite elements is used, this operator necessarily appears in the transient term, and the resulting discretization cannot be treated by using the standard parabolic argument [26, 123]. To overcome these difficulties, a finite volume projection based on the variational formulation of the corresponding finite volume method is first introduced for the Stokes equations. A key argument in the present analysis is to combine this projection and a finite element projection for the Stokes equations without any additional regularity on the exact solution. Furthermore, some results related to the equivalence between the standard L^2-norm and the norm induced by the above mentioned equivalent operator [34] will be used. These techniques, together with the introduction of a duality argument for the derivation of an L^2-error estimate for velocity, will yield convergence rates

© The Author(s), under exclusive license to Springer Nature Switzerland AG 2022
J. Li et al., *Finite Volume Methods for the Incompressible Navier–Stokes Equations*,
SpringerBriefs in Mathematical Methods,
https://doi.org/10.1007/978-3-030-94636-4_4

of optimal order for the present finite volume methods. Compared with the results in [71], the main contribution of this chapter is to establish optimal estimates for the nonstationary Navier–Stokes equations based on the local conservation property. Moreover, we extend the stabilized finite volume methods for the two-dimensional Navier–Stokes equations to the three-dimensional equations. Finally, more research is required on a theoretical analysis of the finite volume methods for the nonstationary Navier–Stokes equations.

The rest of the chapter is organized as follows: In the next section, we introduce the weak formulation of the nonstationary Navier–Stokes equations. Then, in the third section, the Galerkin finite volume methods for these equations are established. Stability and optimal estimates for these methods are obtained in the fourth section. Finally, an L^2 error estimate for velocity is obtained by a technique provided in Chap. 3.

4.2   The Weak Formulation

Let Ω be a bounded domain in $R^d, d = 2, 3$, with a Lipschitz-continuous boundary Γ, satisfying a further condition (A1) as stated in the previous chapter. The nonstationary Navier–Stokes equations are

$$u_t - \mu \Delta u + \nabla p + (u \cdot \nabla)u = f, \quad (x, t) \in \Omega \times [0, T], \qquad (4.1)$$
$$\text{div } u = 0, \quad (x, t) \in \Omega \times [0, T], \qquad (4.2)$$
$$u(x, 0) = u_0(x), \quad x \in \Omega, \qquad (4.3)$$
$$u(x, t) = 0, \quad t \in [0, T], \qquad (4.4)$$

where $u = u(x, t) = (u_1(x, t), u_2(x, t), u_d(x, t))$ represents the velocity vector, $p = p(x, t)$ the pressure, $f = f(x, t)$ the prescribed body force, $\mu > 0$ the viscosity, $T > 0$ the final time, and $u_t = \partial u / \partial t$. Note that the term (div $u)u/2$ is added to ensure the dissipativity of the Navier–Stokes equations [122].

Multiplying Eqs. (4.1) and (4.4) by $v \in X$ and $q \in M$, integrating over Ω, respectively, and applying Green's formula, the mixed variational form of (4.1)–(4.4) is to seek $(u, p) \in X \times M, t > 0$, such that, for all $(v, q) \in X \times M$,

$$(u_t, v) + \mathscr{B}((u, p); (v, q)) + b(u, u, v) = (f, v), \qquad (4.5)$$
$$u(0) = u_0, \qquad (4.6)$$

where $\mathscr{B}((\cdot, \cdot); (\cdot, \cdot))$ is defined in the previous chapter.

We assume the initial value u_0 and the body force f satisfy the following assumption (**A4**):

(A4) The initial velocity $u_0 \in D(A)$ and the body force $f(x, t) \in L^2(0, T; Y)$ are assumed to satisfy

$$\|u_0\|_2 + \left(\int_0^T \left(\|f\|_0^2 + \|f_t\|_0^2 \right) dt \right)^{1/2} \le C.$$

For convenience, we recall the Gronwall Lemma that will be frequently used.

Lemma 4.1 ([116]) *Let $g(t)$, $\ell(t)$, and $\xi(t)$ be three nonnegative functions satisfying, for $s \in [0, T]$,*

$$\xi(s) + G(s) \le c + \int_0^s \ell \, dt + \int_0^s g\xi \, dt,$$

where $G(s)$ is a nonnegative function on $[0, T]$. Then

$$\xi(s) + G(s) \le \left(c + \int_0^s \ell \, dt \right) \exp \left(\int_0^s g \, dt \right). \tag{4.7}$$

The following global existence theorem is provided in ([71], Theorems 2.1 and 2.2):

Theorem 4.1 *With $d = 2$ or 3, assume that $u_0 \in D(A)$, $\sup\limits_{0<t<T} \|f_t\|_0 \le C$, with $\sup\limits_{0<t<T} \|f_t(\cdot, t)\|_0 \le M$, holds for a sufficiently small positive constant M, which ensures that $u \in C([0, T], V)$ and $p \in L^2([0, T], H^1/R)$, with derivatives u_t and $D^2 u$ belonging to $L^2([0, T], L^2)$, all the conditions of (4.1)–(4.4) are satisfied, and the initial condition satisfies the general sense $\|u(\cdot, t) - u_0\|_0 \to 0$ as $t \to 0$. Then, there is at most one such solution for all $t \ge 0$.*

For the purpose of numerical analysis, we need an application and refinement of certain estimates. Moreover, we will directly use the important results related to the regularity theory developed in [71, 92].

Lemma 4.2 *Under the assumptions of* **(A1)**, **(A4)** *and Theorem 4.1, the solution* (u, p) *of* (4.5)–(4.6) *satisfies the following regularities:*

$$\sup_{0<s\leq T} (\|u(s)\|_2^2 + \|p(s)\|_1^2 + \|u_t(s)\|_0^2) + \int_0^T (\|u_t(t)\|_1^2 + \|p_t(t)\|_0^2) dt \leq C,$$

$$\sup_{0<s\leq T} \tau(s)\|u_t(s)\|_1^2 + \int_0^T \tau(t) \left(\|u_t(t)\|_2^2 + \|p_t(t)\|_1^2 + \|u_{tt}(t)\|_0^2\right) dt \leq C,$$

where $\tau(t) = \min\{1, \ t\}$.

Remark ([71]) Although it is unknown whether the conclusion of Theorem 4.1 holds generally for large data in three dimensions, it certainly does in special circumstances; it may sometimes be indicated by experimental observations in cases in which it is still unproven.

Using the above notation, the stabilized finite element formulation of system (4.5) and (4.6) reads: Find $(\bar{u}_h, \bar{p}_h) \in X_h \times M_h$, $t \in [0, T]$, such that, $\forall (v_h, q_h) \in X_h \times M_h$,

$$(\bar{u}_{ht}, v_h) + \mathscr{B}_h((\bar{u}_h, \bar{p}_h); (v_h, q_h)) + b(\bar{u}_h, \bar{u}_h, v_h) = (f, v_h), \tag{4.8}$$

$$u_h(0) = u_{0h}, \tag{4.9}$$

where u_{0h} is some approximation of u_0 in X_h satisfying the approximation property (2.10). To derive error estimates, we define the Stokes projection operators $(\bar{R}_h, \bar{L}_h)$: $X \times M \to X_h \times M_h$ by

$$\mathscr{B}_h((\bar{R}_h(v, q), \bar{L}_h(v, q)); (v_h, q_h)) = B((v, q); (v_h, q_h)) \ \forall (v, q) \in X \times M,$$
$$(v_h, q_h) \in X_h \times M_h, \tag{4.10}$$

which are well defined and satisfy the following approximation property:

Lemma 4.3 *Under the assumption of* (A1), *the projection operator* $(\bar{R}_h, \bar{L}_h)$ *of the finite element method satisfies*

$$\|v - \bar{R}_h(v, q)\|_0 + h \left(\|v - \bar{R}_h(v, q)\|_1 + \|q - \bar{L}_h(v, q)\|_0\right)$$
$$\leq Ch^2(\|v\|_2 + \|q\|_1), \tag{4.11}$$

for all $(v, q) \in D(A) \times H^1(\Omega) \cap M$.

Proof The proof of Lemma 4.3 is classical and can be easily derived from the classical Galerkin finite element method. More details can be found in ([69], Lemma 4.1).

□

Remark The Stokes projection $(\bar{R}_h(u, p), \bar{L}_h(u, p))$ is different from the Stokes projection of the regularized Green's function defined in Chap. 3. Here, the former is available for the stabilized finite element system, which can be degenerated to the latter if the stabilized term $S(\cdot, \cdot) = 0$.

We shall essentially follow the analysis of [92]. We also refer the reader to other presentations, as in [68]. The next optimal estimate is obtained for the stabilized finite element version (4.8) and (4.9).

Theorem 4.2 ([92]) *Under the assumptions of (A1) and (A4), it holds that, for $s \in [0, T]$,*

$$\tau^{1/2}\|u(s) - \bar{u}_h(s)\|_0 + h(\|u(s) - \bar{u}_h(s)\|_1 + \tau^{1/2}\|p(s) - \bar{p}_h(s)\|_0) \leq \kappa h^2. \tag{4.12}$$

4.3 Galerkin FV Approximation

Based on the previous results, the corresponding discrete finite volume variational formulation of (4.5) and (4.6) for the Navier–Stokes equations is recast: Find $(u_h, p_h) \in X_h \times M_h$ such that, $\forall (v_h, q_h) \in X_h \times M_h$,

$$(u_{th}, v_h^*) + \mathscr{C}_h((u_h, p_h), (v_h, q_h)) + b(u_h; u_h, v_h^*) = (f, v_h^*), \tag{4.13}$$
$$u_h(x, 0) = P_h u_0(x), \tag{4.14}$$

where the approximation $u_h(x, 0) \in X_h$ of the initial value u_0 is given as follows:

$$(P_h u_0 - u_0, v_h^*) = 0,$$

which satisfies (2.10).

In this chapter, optimal results can be obtained by reasonable regularity and useful techniques due to the definition of the finite volume methods and the lower order error $O(h)$ between the test functions of the these methods and those of finite element methods.

Lemma 4.4 *The mapping Γ_h is self-adjoint with respect to the L^2-inner product:*

$$(u_h, v_h^*) = (u_h^*, v_h) \quad \forall u_h, \ v_h \in X_h. \tag{4.15}$$

In addition, the norm

$$\|u_h\|_0 = (u_h, u_h^*)^{1/2}$$

is equivalent to the usual L^2-norm

$$c_{12}\|u_h\|_0 \leq \|u_h\|_0 \leq c_{13}\|u_h\|_0, \tag{4.16}$$

where the constants $c_{12} > 0$ and $c_{13} > 0$ are independent of h. In particular, if $u_h(\cdot, t) \in X_h$ and $v_h \in X_h$, $t \in [0, T]$, it holds

$$(u_{th}, v_h^*) = (u_{th}^*, v_h). \tag{4.17}$$

Finally, if $v_h = u_h$, it holds

$$(u_{th}, u_h^*) = (u_{th}^*, u_h) = \frac{1}{2}\frac{d}{dt}\|u_h\|_0^2. \tag{4.18}$$

Proof Results (4.15) and (4.16) can be found in [34]. For completeness, we prove (4.17). Denote the vertices of an element K by P_1, P_2, ..., P_{d+1}. Let φ_i be a basis function in K and Q_i, $i = 1, 2, \ldots, d + 1$, be a quadrilateral or polyhedron (see Fig. 2.1). For fixed t, any functions $u_h(\cdot, t)$ and $v_h(\cdot, t)$ have the unique representations

$$u_h(x, t)|_K = \sum_{i=1}^{d+1} u_i(t)\varphi_i(x), \quad v_h(x, t)|_K = \sum_{j=1}^{d+1} v_j(t)\varphi_j(x), \quad x \in \Omega,$$

$$\varphi_i(x) \in X_h, \quad (4.19)$$

where $u_i(t)$ and $v_j(t)$, $i, j = 1, 2, \ldots, d + 1$, are the values of u_h and v_h at the node P_i, respectively.

Using the fact that

$$v_h^* = \sum_{j=1}^{d+1} v_j(t)\chi_j(x)$$

yields

$$(u_{th}, v_h^*)_K = \sum_{K \in K_h} \int_K \left(\sum_{i=1}^{d+1} \frac{du_i(t)}{dt} \varphi_i \right) v_h^* \, dx$$

$$= \sum_{K \in K_h} \sum_{j=1}^{d+1} v_j(t) \int_{Q_j} \sum_{i=1}^{d+1} \frac{du_i(t)}{dt} \varphi_i \, dx$$

$$= \sum_{K \in K_h} \sum_{i=1}^{d+1} \sum_{j=1}^{d+1} \frac{du_i(t)}{dt} v_j(t) \int_{Q_j} \varphi_i \, dx.$$

Since the term $\frac{du_i(t)}{dt}$ is independent of $x \in \Omega$, we can extract it from the integral over Q_j. Use $\int_{Q_j} \varphi_i dx = \int_{Q_i} \varphi_j dx$ to obtain

$$(u_{th}, v_h^*)_K = \sum_{K \in K_h} \sum_{i=1}^{d+1} \sum_{j=1}^{d+1} \frac{du_i(t)}{dt} v_j(t) \int_{Q_i} \varphi_j \, dx$$

$$= \sum_{K \in K_h} \sum_{i=1}^{d+1} \int_{Q_i} \sum_{j=1}^{d+1} v_j(t) \varphi_j (\partial_t u_i(t))^* \, dx$$

$$= (u_{th}^*, v_h)_K, \tag{4.20}$$

where we used $\int_{e_i} \varphi_j dx = \int_{e_j} \varphi_i dx$ [34]. In particular, if $v_h = u_h$, we have

$$(u_{th}, u_h^*) = (u_{th}^*, u_h) = \frac{1}{2} \frac{d}{dt} \|u_h\|_0^2.$$

Consequently, (4.17) and (4.18) are proven. □

To derive error estimates for the finite volume solution (u_h, p_h), we define a projection operator $(R_h, L_h) : X \times M \to X_h \times M_h$ by

$$\mathscr{C}_h((u - R_h(u, p), p - L_h(u, p)); (v_h, q_h)) = S(p, q_h) \; \forall (u, p) \in X \times M,$$
$$(v_h, q_h) \in X_h \times M_h, \tag{4.21}$$

which is well defined by Theorem 2.1. Also, it satisfies the following stability and approximation properties ([85], Lemma 4.5):

Lemma 4.5 *Under the assumption of (A1), the projection operator (R_h, L_h) satisfies*

$$\|R_h(u, p)\|_1 + \|L_h(u, p)\|_0 \le C(\|u\|_1 + \|p\|_0), \qquad (4.22)$$
$$\|u - R_h(u, p)\|_1 + \|p - L_h(u, p)\|_0 \le Ch(\|u\|_2 + \|p\|_1), \quad (4.23)$$

for all $(u, p) \in D(A) \times H^1(\Omega) \cap M$.

Proof The stability property (4.22) can easily be verified. It remains to prove the approximation property (4.23). Using simple calculations and setting $E = u - I_h u$ and $e = I_h u - R_h(u, p)$, we see that

$$A(e, v_h^*) + D(v_h^*, \eta) + d(e, q_h) + S(\eta, q_h) = -A(E, v_h^*) - D(v_h^*, p - J_h p)$$
$$-d(E, q_h) - S(p - J_h p, q_h) + S(p, q_h). \quad (4.24)$$

Using the interpolation property (2.10), (2.16) and (2.17) gives

$$|d(E, q_h) + S(p - J_h p, q_h) - S(p, q_h)| \le Ch(\|u\|_2 + \|p\|_1)\|q_h\|_0. \quad (4.25)$$

Then, we deduce from Green's formula, the Cauchy-Schwarz inequality, the interpolation property (2.10), and estimate (2.24) between the functions of the finite element and finite volume methods that

$$
\begin{aligned}
|D(v_h^*, p - J_h p)| &= \left| \sum_K (\nabla(p - J_h p), v_h^*)_K \right| \\
&\le \left| \sum_K (\nabla(p - J_h p), v_h^* - v_h)_K \right| + |(\nabla(p - J_h p), v_h)| \\
&\le \sum_K \|\nabla(p - J_h p)\|_{0,K} \|v_h^* - v_h\|_{0,K} + |(\text{div}v_h, p - J_h p)| \\
&\le Ch\|p\|_1 \|v_h\|_1. \quad (4.26)
\end{aligned}
$$

Since

$$A(E, v_h^*) = -\sum_{j \in N_h} \sum_{K \cap V_j} \int_{\partial V_j \cap K} \frac{\partial E}{\partial n} v_h^* ds$$

$$= \sum_K \int_{\partial K} \frac{\partial E}{\partial n} v_h^* ds - \sum_K (\Delta E, v_h^*)_K$$

$$= \sum_K \int_{\partial K} \left(\frac{\partial E}{\partial n} - \frac{\partial \hat{\pi}_h E}{\partial n} \right) (v_h^* - v_h) ds - \sum_K (\Delta E, v_h^* - v_h)_K$$

$$- \sum_K (\nabla E, \nabla v_h)_K,$$

a similar argument yields

$$|A(E, v_h^*)| \le \sum_K \|\nabla(E - \hat{\pi}_h E)\|_{0,\partial K} \|v_h^* - v_h\|_{0,\partial K}$$

$$+ \sum_K \|E\|_{2,K} \|v_h - v_h^*\|_{0,K} + c\|E\|_1 \|v_h\|_1$$

$$\le Ch\|u\|_2 \|v_h\|_1. \tag{4.27}$$

Using all these inequalities (4.24)–(4.27), we find that

$$\|I_h u - R_h(u, p)\|_1 + \|J_h p - L_h(u, p)\|_0$$

$$\le \frac{1}{\beta^*} \sup_{(v_h, q_h) \in X_h \times M_h} \frac{|\mathscr{C}_h((I_h u - R_h(u, p), J_h p - L_h(u, p)); (v_h, q_h))|}{\|v_h\|_1 + \|q_h\|_0}$$

$$\le Ch(\|u\|_2 + \|p\|_1). \tag{4.28}$$

Finally, combining (4.28) and (2.10) gives (4.23). □

So far, we rely on two kinds of the Stokes projections, based on a Galerkin system and a Petrov–Galerkin system. The first Stokes projection defined in (4.10) has optimal convergence results and leads to an optimal estimate for the finite element discretization of the nonstationary Navier–Stokes equations. For the second one defined in (4.21), we cannot obtain an optimal estimate for $\|u - R_h(u, p)\|_0$ because of a lower order error between the test functions of the finite element and finite volume methods. However, it is worth to mention that both projections complement perfectly each other to obtain optimal results for the finite volume methods of the nonstationary Navier–Stokes equations without any additional regularity on the exact solution.

4.4 Stability and Error Analysis

In this section, the main focus is to provide a stability and error analysis for the stabilized finite volume methods based on the relationship between the finite element methods and the finite volume methods and some additional analytical techniques.

Lemma 4.6 *Assume that the initial value u_0 small enough, (A1), (A4) holds and*

$$\frac{4^3 C_4^2 C_{10}^2 \gamma^2}{\mu^4} \left\{ \left(C_4^2 \int_0^s \|f_t\|_0^2 dt + C_{15}^2 \right) e^T + \|f\|_0^2 \right\} < 1, \quad (4.29)$$

if

$$0 < \frac{2 C_2 C_3 h^{1/2}}{C_{10}} < 1, \quad (4.30)$$

for sufficiently small $h > 0$, it holds that, for $s \in [0, T]$,

$$\|u_h(s)\|_1 < \frac{\mu}{4 C_{10}}. \quad (4.31)$$

Proof Based on the previous finite volume algorithm, the initial value produced by the finite volume solution of the Stokes equations can obviously be bounded by some positive constant independent of h [84, 93].   □

Observe that under the assumption on u_0, the continuity of the solution guarantees that $\nabla u_h(t)$ will stay as small as we wish in an interval $[0, \bar{T}_m]$, where $0 < \bar{T}_m \leq T_m$ depends upon the smallness condition we prescribe. Here, we will estimate:

$$\|u_h(s)\|_1 < \frac{\mu}{4 C_{10}}, \quad (4.32)$$

for all $s \in [0, T]$. By the contradiction assumption, there exists a $T^* \in (0, T]$ such that

$$\|u_h(s)\|_1 < \frac{\mu}{4 C_{10}}, \quad s \in [0, T^*) \quad (4.33)$$

and

$$\|u_h(T^*)\|_1 = \frac{\mu}{4 C_{10}}. \quad (4.34)$$

Taking $(v_h, q_h) = (u_h, p_h)$ in (4.13) and (4.14), using assumption (4.30), (4.33), the inverse inequality (2.13), and applying the bound of the trilinear term, we have

$$b(u_h, u_h, u_h^*) = b(u_h, u_h, u_h^* - u_h)$$

$$\leq \|u_h\|_{L^\infty} \|u_h\|_1 \|u_h^* - u_h\|_0 + \frac{\sqrt{d}}{2} \|u_h\|_{L^\infty} \|u_h\|_1 \|u_h^* - u_h\|_0$$

$$\leq 2C_2 C_3 h^{1/2} \|u_h\|_1^3$$

$$\leq \frac{\mu}{4} \|u_h\|_1^2. \tag{4.35}$$

For sufficiently small h, we see that

$$\frac{1}{2}\frac{d}{dt} \|u_h\|_0^2 + \frac{3\mu}{4} \|u_h\|_1^2 + S(p_h, p_h) \leq C_4 \|f\|_0 \|u_h\|_0$$

$$\leq \frac{C_4^2}{2} \|f\|_0^2 + \frac{1}{2} \|u_h\|_0^2. \tag{4.36}$$

Integrating the above inequality from 0 to $s \in [0, T^*]$ with respect to time, applying the Gronwall inequality, and noting that

$$\|u_h(0)\|_0 \leq \|P_h u_0\|_0 \leq C \|u_0\|_1 \leq C_{14},$$

we obtain

$$\|u_h(s)\|_0^2 + \int_0^s (\mu \|u_h\|_1^2 + S(p_h, p_h)) dt \leq \left(C_4^2 \int_0^s \|f\|_0^2 dt + C_{14}^2 \right) e^{T^*}. \tag{4.37}$$

Substituting $(v_h, q_h) = (u_h, p_h)$ into (4.13)–(4.14), applying the same approach as for (4.35), and using the Young inequality, we have

$$\mu \|u_h\|_1^2 + S(p_h, p_h) = -(u_{th}, u_h^*) - b(u_h, u_h, u_h^*) + (f, u_h^*)$$

$$\leq C_4 \gamma \|u_{th}\|_0 \|u_h\|_1 + \frac{\mu}{4} \|u_h\|_1^2 + C_4 \gamma \|f\|_0 \|u_h\|_1$$

$$\leq \frac{2C_4^2 \gamma^2 \|u_{th}\|_0^2}{\mu} + \frac{2C_4^2 \gamma^2}{\mu} \|f\|_0^2 + \frac{\mu}{2} \|u_h\|_1^2,$$

which is

$$\mu \|u_h\|_1^2 \leq \frac{4C_4^2 \gamma^2}{\mu} (\|u_{th}\|_0^2 + \|f\|_0^2). \tag{4.38}$$

Furthermore, we need to prove a bound for $\|u_{th}\|_0$ in the last term. First, we differentiate Eq. (4.13) with respect to time and take $(v_h, q_h) = (u_{th}, p_{th})$ to obtain

$$\frac{1}{2}\frac{d}{dt} \|u_{th}\|_0^2 + \mu \|u_{th}\|_1^2 + S(p_{th}, p_{th})$$

$$= -b(u_{th}, u_h, u_{th}^*) - b(u_h, u_{th}, u_{th}^*) + (f_t, u_{th}^*). \tag{4.39}$$

Here, using the inverse inequality (2.13), we see that

$$
\begin{aligned}
b(u_{th}, u_h, u_{th}^*) &= b(u_{th}, u_h, u_{th}^* - u_{th}) + b(u_{th}, u_h, u_{th}) \\
&\leq \|u_{th}\|_{L^\infty} \|u_h\|_1 \|u_{th}^* - u_{th}\|_0 + \frac{\sqrt{d}}{2} \|u_{th}\|_1 \|u_h\|_{L^\infty} \|u_{th}^* - u_{th}\|_0 \\
&\quad + C_{10} \|u_{th}\|_1^2 \|u_h\|_1 \\
&\leq 2 C_2 C_3 h^{1/2} \|u_h\|_1 \|u_{th}\|_1^2 + C_{10} \|u_{th}\|_1^2 \|u_h\|_1.
\end{aligned}
$$

Similarly, we see that

$$
\begin{aligned}
b(u_h, u_{th}, u_{th}^*) &= b(u_h, u_{th}, u_{th}^* - u_{th}) \\
&\leq 2 C_2 C_3 h^{1/2} \|u_h\|_1 \|u_{th}\|_1^2.
\end{aligned}
\tag{4.40}
$$

By (4.30) for sufficiently small h, the trilinear term in (4.39) can be bounded by the following inequality:

$$
\begin{aligned}
b(u_{th}, u_h, u_{th}^*) &+ b(u_h, u_{th}, u_{th}^*) \\
&\leq 4 C_2 C_3 h^{1/2} \|u_h\|_1 \|u_{th}\|_1^2 + C_{10} \|u_{th}\|_1^2 \|u_h\|_1 \\
&\leq \frac{\mu}{2} \|u_{th}\|_1^2 + C_{10} \|u_{th}\|_1^2 \|u_h\|_1,
\end{aligned}
\tag{4.41}
$$

which, together with (4.39), yields that

$$
\begin{aligned}
\frac{1}{2} \frac{d}{dt} \|u_{th}\|_0^2 + \mu \|u_{th}\|_1^2 + S(p_{th}, p_{th}) &= C_{10} \|u_{th}\|_1^2 \|u_h\|_1 + C_4 \|f_t\|_0 \|u_{th}\|_0 \\
&= \mu \|u_{th}\|_1^2 + \frac{C_4^2}{2} \|f_t\|_0^2 + \frac{1}{2} \|u_{th}\|_0^2.
\end{aligned}
\tag{4.42}
$$

By simple calculations and applying the Gronwall lemma, we find that

$$
\begin{aligned}
\|u_{th}\|_0^2 + \int_0^s (\mu \|u_{th}\|_1^2 + S(p_{th}, p_{th})) dt \\
\leq \left(C_4^2 \int_0^s \|f_t\|_0^2 dt + \|u_{th}(0)\|_0^2 \right) e^{T^*}.
\end{aligned}
\tag{4.43}
$$

Then, taking $t = 0$ in (4.13), we can easily get the following bound:

$$
\|u_{th}(0)\|_0 \leq C_{15},
\tag{4.44}
$$

where C_{15} is dependent of the data (Ω, μ, f).

Substituting (4.43)–(4.44) into (4.38), we have

$$
\mu \|u_h\|_1^2 \leq \frac{4 C_4^2 \gamma^2}{\mu} \left\{ \left(C_4^2 \int_0^s \|f_t\|_0^2 dt + C_{15}^2 \right) e^{T^*} + \|f\|_0^2 \right\}.
\tag{4.45}
$$

Using (4.29), we can obtain, for all $s \in [0, T^*]$,

$$\|u_h(s)\|_1 < \frac{\mu}{4C_{10}} \quad \forall s \in [0, T^*], \tag{4.46}$$

which implies

$$\|u_h(T^*)\|_1 < \frac{\mu}{4C_{10}}. \tag{4.47}$$

Obviously, the findings appear to contradict assumptions (4.34). Then, using the same approach as above for $s \in [0, T]$, we obtain (4.31).

Lemma 4.7 *Under the assumptions of (A1), (A4), and Lemma 4.6, it holds that, for $s \in [0, T]$,*

$$\|u(s) - u_h(s)\|_0^2 + \int_0^s \left(\mu \|u - u_h\|_1^2 + S(p - p_h, p - p_h) \right) dt \leq Ch^2. \tag{4.48}$$

Proof Multiplying (4.1) and (4.2) by v_h^* and q_h, integrating over V and K, summing over all elements, respectively, and using (4.13) and (4.14), we see that

$$(u_t - u_{th}, v_h^*) + \mathscr{C}_h((u - u_h, p - p_h); (v_h, q_h)) + b(u - u_h, u, v_h) + b(u_h, u - u_h, v_h)$$
$$+ b(u - u_h, u, v_h^* - v_h) + b(u_h, u - u_h, v_h^* - v_h) = S(p, q_h). \tag{4.49}$$

Setting $(e_h, \eta_h) = (R_h(u, p) - u_h, L_h(u, p) - p_h)$ and $E = u - R_h(u, p)$, using the equivalence Lemma 2.3 and (3.8), and taking $(v_h, q_h) = (e_h, \eta_h)$ in (4.49), we have

$$\frac{1}{2} \frac{d}{dt} \|e_h\|_0^2 + \mu \|e_h\|_1^2 + S(\eta_h, \eta_h) + b(E + e_h, u, e_h) + b(u_h, E, e_h)$$
$$+ b(E + e_h, u, e_h^* - e_h) + b(u_h, E + e_h, e_h^* - e_h)$$
$$= S(p, q_h) - (E_t, e_h^*). \tag{4.50}$$

Using Lemma 4.5, (2.25), and the Young inequality, we see that

$$|(E_t, e_h^*)| \leq C \|E_t\|_0 \|e_h\|_0 \leq Ch^2 (\|u_t\|_1^2 + \|p_t\|_0^2) + \|e_h\|_0^2.$$

As for the stabilization term, using (2.17) gives

$$|S(p, \eta_h)| \leq \|p - \Pi_h p\|_0 \|\eta_h - \Pi_h \eta_h\|_0 \leq \frac{1}{2} S(\eta_h, \eta_h) + Ch^2 \|p\|_1^2.$$

By the bounds of the trilinear term, we have

$$|b(E, u, e_h)| \leq C\|E\|_1\|u\|_1\|e_h\|_1 \leq \frac{\mu}{16}\|e_h\|_1^2 + C\|u\|_1^2\|E\|_1^2,$$

$$|b(u_h, E, e_h)| \leq C\|u_h\|_1\|E\|_1\|e_h\|_1 \leq \frac{\mu}{16}\|e_h\|_1^2 + C\|u_h\|_1^2\|E\|_1^2,$$

$$|b(e_h, u, e_h)| \leq C\|e_h\|_1\|u\|_2\|e_h\|_0 \leq \frac{\mu}{16}\|e_h\|_1^2 + C\|u\|_2^2\|e_h\|_0^2.$$

Similarly, we see that

$$|b(E, u, e_h^* - e_h)| \leq C\|E\|_1\|u\|_2\|e_h^* - e_h\|_0 \leq \tfrac{\mu}{16}\|e_h\|_1^2 + Ch^2\|u\|_2^2\|E\|_1^2,$$

$$|b(e_h, u, e_h^* - e_h)| \leq C\|u\|_2\|e_h\|_1\|e_h^* - e_h\|_0 \leq \tfrac{\mu}{16}\|e_h\|_1^2 + C\|u\|_2^2\|e_h\|_0^2.$$

Using the L^∞ inequality (2.13) and the Young inequality yields

$$|b(u_h, E, e_h^* - e_h)|$$

$$\leq C\|u_h\|_{L^\infty}\|E\|_1\|e_h^* - e_h\|_0 + \frac{\sqrt{d}}{2}\|u_h\|_1\|E\|_{L^4}\|e_h^* - e_h\|_{L^4}$$

$$\leq Ch^{1/2}\|u_h\|_1\|E\|_1\|e_h\|_1 + C\|u_h\|_1\|E\|_0^{d/4}\|E\|_1^{1-d/4}\|e_h^* - e_h\|_0^{d/4}\|e_h^* - e_h\|_0^{1-d/4}$$

$$\leq Ch^{1/2}\|u_h\|_1\|E\|_1\|e_h\|_1$$

$$\leq \frac{\mu}{16}\|e_h\|_1^2 + C\|u_h\|_2^2\|E\|_0^2.$$

$\square$

Also, applying the estimates in the L^∞- and L^4-norms and the Hölder inequality, we have

$$|b(u_h, e_h, e_h^* - e_h)| \leq C(\|u_h\|_{L^\infty}\|\nabla e_h\|_0\|e_h - e_h^*\|_0 + \|\nabla e_h\|_0\|e_h\|_{L^4}\|e_h - e_h^*\|_{L^4})$$

$$\leq C_2C_3h^{1/2}\|u_h\|_1\|e_h\|_1^2 + Ch^{\frac{d}{4}}\|u_h\|_1\|e\|_0^{\frac{d}{4}}\|e\|_1^{1-\frac{d}{4}}\|e\|_1$$

$$\leq 2C_2C_3h^{1/2}\|u_h\|_1\|e_h\|_1^2$$

$$\leq \frac{C_2C_3h^{1/2}\mu}{4C_{10}}\|e_h\|_1^2$$

$$\leq \frac{\mu}{8}\|e_h\|_1^2, \tag{4.51}$$

since

$$Ch^{\frac{d}{4}}\|u_h\|_1\|e\|_0^{\frac{d}{4}}\|e\|_1^{1-\frac{d}{4}}\|e\|_1 \leq Ch^{1/2}\|u_h\|_1\|e_h\|_1^2,$$

for sufficiently small $h > 0$.

Now, absorbing (4.51) by the second term $\mu\|e_h\|_1^2$ in the left-hand side in (4.50), substituting these inequalities into (4.50), and using (A4), we find

$$\frac{d}{dt}\|e_h\|_0^2 + \mu\|e_h\|_1^2 + S(\eta_h, \eta_h) \leq C\Big\{h^2(\|u\|_2^2 + \|u_h\|_1^2 + \|u_t\|_1^2 + \|p_t\|_0^2)$$

$$+(1 + \|u\|_2^2)\|e_h\|_0^2\Big\}.$$

Therefore, integrating the above inequality from 0 to $s \in [0, T]$, we deduce from the Schwarz inequality, the Gronwall Lemma 4.1 the regularity Lemma 4.2, and (4.18) that

$$\|e_h\|_0^2 + \int_0^s \left(\mu\|e_h\|_1^2 + S(\eta_h, \eta_h)\right) dt \leq Ch^2, \tag{4.52}$$

which, together with (4.23), yields

$$\|u - u_h\|_0^2 + \int_0^s \left(\mu\|u - u_h\|_1^2 + S(p - p_h, p - p_h)\right) dt \leq Ch^2. \tag{4.53}$$

Thus the desired result (4.48) follows.

Lemma 4.8 *Under the assumptions of (A1), (A4), and Lemma 4.6, it holds that, for $s \in [0, T]$,*

$$\int_0^s \|u_{th}\|_0^2 dt \leq C. \tag{4.54}$$

Proof Differentiating the term $d(u_h, q_h) + S(p_h, q_h)$ with respect to time t, taking $(v_h, q_h) = (u_{th}, p_h)$ in (4.13) and (4.14), and using the equivalence Lemma 2.3, we see that

$$\|u_{th}\|_0^2 + \frac{1}{2}\frac{d}{dt}(\mu\|u_h\|_1^2 + S(p_h, p_h)) + b(u_h, u - u_h, u_{th} - u_{th}^*)$$
$$-b(u - u_h, u, u_{th}^*) - b(u_h, u - u_h, u_{th}) + b(u, u, u_{th}^*) = (f, u_{th}^*). \tag{4.55}$$

Applying the estimates of the trilinear terms and the Young inequality leads to

$$|b(u - u_h, u, u_{th}^*)| \leq C\|u - u_h\|_1\|u\|_2\|u_{th}^*\|_0$$
$$\leq \frac{1}{10}\|u_{th}\|_0^2 + C\|u\|_2^2\|u - u_h\|_1^2,$$

$$|b(u, u, u_{th}^* - u_{th}))| \leq C\|u\|_2\|u\|_1\|u_{th}\|_0$$
$$\leq \frac{1}{10}\|u_{th}\|_0^2 + C\|u\|_1^2\|u\|_2^2.$$

A combination of the second inequality of (3.7) and the Cauchy-Schwarz inequality implies that

$$
\begin{aligned}
&|b(u_h, u - u_h, u_{th})| \\
&\leq \|u_h\|_{L^4}\|u - u_h\|_1\|u_{th}\|_{L^4} + \|\nabla u_h\|_0\|u - u_h\|_{L^4}\|u_{th}\|_{L^4} \\
&\leq Ch^{\frac{d}{4}-1}(\|u_h\|_0^{\frac{d}{4}}\|u_h\|_1^{1-\frac{d}{4}}\|u - u_h\|_1 + \|u_h\|_1\|u - u_h\|_0^{\frac{d}{4}}\|u - u_h\|_1^{1-\frac{d}{4}})\|u_{th}\|_0 \\
&\leq \frac{1}{10}\|u_{th}\|_0^2 + Ch^{\frac{d}{2}-2}\left(\|u_h\|_0^{\frac{d}{2}}\|u_h\|_1^{2-\frac{d}{2}}\|u - u_h\|_1^2 + \|u_h\|_1^2\|u - u_h\|_0^{\frac{d}{2}}\|u - u_h\|_1^{2-\frac{d}{2}}\right) \\
&\leq \frac{1}{10}\|u_{th}\|_0^2 + Ch^{\frac{d}{2}-2}\|u_h\|_1^2\|u - u_h\|_1^2.
\end{aligned}
\tag{4.56}
$$

Similarly, we see that

$$
|b(u_h, u - u_h, u_{th} - u_{th}^*)| \leq \frac{1}{10}\|u_{th}\|_0^2 + Ch^{\frac{d}{2}-2}\|u_h\|_1^2\|u - u_h\|_1^2. \tag{4.57}
$$

In addition, a simple computation shows that

$$
|(f, u_{th}^*)| \leq \|f\|_0\|u_{th}\|_0 \leq \frac{1}{10}\|u_{th}\|_0^2 + \|f\|_0^2.
$$

Now, combining all these inequalities and (4.55) yields

$$
\begin{aligned}
\|u_{th}\|_0^2 + \tfrac{d}{dt}(\mu\|u_h\|_1^2 + S(p_h, p_h)) \leq C\Big\{&\|u\|_2^2\|u - u_h\|_1^2 + \|u\|_1^2\|u\|_2^2 + \|f\|_0^2 \\
&+ h^{\frac{d}{2}-2}\|u_h\|_1^2\|u - u_h\|_1^2\Big\}.
\end{aligned}
\tag{4.58}
$$

Then, integrating (4.58) from 0 to s with respect to time, using (4.21), and noting that

$$
\mu\|u_h(0)\|_1^2 + S(p_h(0), p_h(0)) \leq C(\|u\|_1^2 + \|p\|_0^2),
$$

we deduce from (4.31), (4.48), and (A4) that

$$
\begin{aligned}
\mu\|u_h\|_1^2 + S(p_h, p_h) + \int_0^s \|u_{th}\|_0^2 dt \leq{}& C(\|u\|_1^2 + \|p\|_0^2) + C\Big\{\|u\|_2^2\int_0^s\|u - u_h\|_1^2 dt \\
&+ \int_0^s(\|u\|_1^2\|u\|_2^2 + \|f\|_0^2)dt + h^{\frac{d}{2}-2}\|u_h\|_1^2\int_0^s\|u - u_h\|_1^2 dt\Big\} \leq C,
\end{aligned}
$$

which complete the proof of (4.54).   □

Remark This lemma is consistent with the findings of Theorems 2.1 and 2.2 in [71].

Lemma 4.9 *Under the assumptions of (A1), (A4), and Lemma 4.6, it holds that, for $s \in [0, T]$,*

$$\mu \|u(s) - u_h(s)\|_1^2 + \int_0^s \|u_t - u_{th}\|_0^2 \, dt \le Ch^2. \tag{4.59}$$

Proof Differentiating the term $d(u - u_h, q_h) + S(p - p_h, q_h)$ with respect to time, taking $(v_h, q_h) = (e_{th}, \eta_h) = (R_{th}(u, p) - u_{th}, L_h(u, p) - p_h)$ in (4.49), and using Lemmas 2.3 and 4.4, we see that

$$\|e_{th}\|_0^2 + \frac{1}{2}\frac{d}{dt}\left(\mu\|e_h\|_1^2 + S(\eta_h, \eta_h)\right) + b(u - u_h, u, e_{th}) + b(u, u - u_h, e_{th})$$
$$-b(u - u_h, u - u_h, e_{th}) + b(u, u - u_h, e_{th}^* - e_{th}) + b(u - u_h, u, e_{th}^* - e_{th})$$
$$-b(u - u_h, u - u_h, e_{th}^* - e_{th}) = -(E_t, e_{th}^*). \tag{4.60}$$

Applying the estimates of the trilinear term, we have

$$|b(u - u_h, u, e_{th}) + b(u, u - u_h, e_{th})| \le C\|u\|_2\|u - u_h\|_1\|e_{th}\|_0$$
$$\le \frac{1}{10}\|e_{th}\|_0^2 + C\|u\|_2^2\|u - u_h\|_1^2,$$
$$|b(u, u - u_h, e_{th}^* - e_{th}) + b(u - u_h, u, e_{th}^* - e_{th})| \le C\|u\|_2\|u - u_h\|_1\|e_{th}\|_0$$
$$\le \frac{1}{10}\|e_{th}\|_0^2 + C\|u\|_2^2\|u - u_h\|_1^2. \tag{4.61}$$

Using the Hölder inequality, the second inequality of (3.7), the inverse inequality (2.12) and the Young inequality gives

$$|b(u_h - u, u - u_h, e_{th})| \le \|u - u_h\|_{L^4}\|u - u_h\|_1\|e_{th}\|_{L^4}$$
$$\le C\|u - u_h\|_0^{\frac{d}{4}}\|u - u_h\|_1^{2-\frac{d}{4}}\|e_{th}\|_0^{\frac{d}{4}}\|e_{th}\|_1^{1-\frac{d}{4}}$$
$$\le Ch^{\frac{d}{4}-1}\|u - u_h\|_0^{\frac{d}{4}}\|u - u_h\|_1^{2-\frac{d}{4}}\|e_{th}\|_0$$
$$\le \frac{1}{10}\|e_{th}\|_0^2 + Ch^{\frac{d}{2}-2}\|u - u_h\|_0^{\frac{d}{2}}\|u - u_h\|_1^{4-\frac{d}{2}}.$$

Similarly, we can bound the following inequality:

$$|b(u - u_h, u - u_h, e_{th} - e_{th}^*)| \le \frac{1}{10}\|e_{th}\|_0^2 + Ch^{\frac{d}{2}-2}\|u - u_h\|_0^{\frac{d}{2}}\|u - u_h\|_1^{4-\frac{d}{2}}.$$

Using Lemma 4.3 and the Young inequality leads to

$$|(E_t, e_{th}^*)| \le C\|E_t\|_0\|e_{th}\|_0 \le \frac{1}{10}\|e_{th}\|_0^2 + Ch^2(\|u_t\|_1^2 + \|p_t\|_0^2).$$

Hence, combining these inequalities and (4.60) yields

$$\|e_{th}\|_0^2 + \frac{d}{dt}\left(\mu\|e_h\|_1^2 + S(\eta_h, \eta_h)\right) \le C\Big\{\|u\|_2^2\|u - u_h\|_1^2$$

$$+h^2(\|u_t\|_1^2 + \|p_t\|_0^2) + Ch^{\frac{d}{2}-2}\|u - u_h\|_0^{\frac{d}{2}}\|u - u_h\|_1^{4-\frac{d}{2}}\Big\}.$$

Integrating the above inequality from 0 to s and using (4.48) and $(A4)$, we see that

$$\mu\|e_h(s)\|_1^2 + \int_0^s \|e_{th}\|_0^2 dt \le C\|u\|_2^2 \int_0^s \|u - u_h\|_1^2 ds$$

$$+Ch^2 \int_0^s (\|u_t\|_1^2 + \|p_t\|_0^2)dt + Ch^{d-2}\int_0^s \|u - u_h\|_1^{4-\frac{d}{2}}dt) + \mu\|e(0)\|_0^2, \quad (4.62)$$

which, together with Lemmas 4.2 and 4.5, yields the desired result.   □

Lemma 4.10 *Under the assumptions of (A1), (A4), and Lemma 4.6, it holds that, for $s \in [0, T]$,*

$$\|u_{th}(s)\|_0^2 + \int_0^s \left(\mu\|u_{th}\|_1^2 + S(p_{th}, p_{th})\right) dt \le C, \quad (4.63)$$

$$\tau(s)\left(\mu\|u_{th}(s)\|_1^2 + S(p_{th}(s), p_{th}(s))\right) + \int_0^s \tau(t)\|u_{tth}\|_0^2 dt \le C. \quad (4.64)$$

Proof By differentiating (4.13) and (4.14) with respect to time, it follows that

$$(u_{tth}, v_h^*) + \mathscr{C}_h((u_{th}, p_{th}); (v_h, q_h)) + b(u_{th}, u_h, v_h^*) + b(u_h, u_{th}, v_h^*)$$
$$= (f_t, v_h^*) \quad \forall\, (v_h, q_h) \in X_h \times M_h. \quad (4.65)$$

Substituting $(v_h, q_h) = 2(u_{th}, p_{th})$ into (4.65) and using the equivalence Lemma 2.3 and the skew-symmetry property (3.8), we have

$$\frac{d}{dt}\|u_{th}\|_0^2 + 2\mu\|u_{th}\|_1^2 + 2S(p_{th}, p_{th}) + 2b(u_{th}, u_h, u_{th}^* - u_{th}) + 2b(u_{th}, u_h, u_{th})$$

$$+2b(u_h, u_{th}, u_{th}^* - u_{th}) \le \|u_{th}\|_0^2 + C\|f_t\|_0^2. \quad (4.66)$$

Applying (3.7), the Hölder inequality, and the Young inequality twice, we find that

$$2|b(u_{th}, u_h, u_{th})|$$
$$\leq C(\|u_{th}\|_{L^4}\|u_h\|_1\|u_{th}\|_{L^4} + \|u_{th}\|_1\|u_h\|_{L^4}\|u_{th}\|_{L^4}$$
$$\leq C(\|u_{th}\|_0^{d/2}\|u_{th}\|_1^{2-d/2}\|u_h\|_1 + \|u_{th}\|_1\|u_h\|_0^{\frac{d}{4}}\|u_h\|_1^{1-\frac{d}{4}}\|u_{th}\|_0^{\frac{d}{4}}\|u_{th}\|_1^{1-\frac{d}{4}}$$
$$\leq C(\|u_{th}\|_0\|u_h\|_1\|u_{th}\|_1 + \|u_h\|_0^{1/2}\|u_h\|_1^{1/2}\|u_{th}\|_0^{1/2}\|u_{th}\|_1^{3/2})$$
$$\leq \frac{\mu}{2}\|u_{th}\|_1^2 + C(1 + \|u_h\|_0^2)\|u_h\|_1^2\|u_{th}\|_0^2. \tag{4.67}$$

Obviously, noting that $\|u_{th} - u_{th}^*\|_i \leq C\|u_{th}\|_i$, we estimate the trilinear term with the available bounds as follows:

$$2|b(u_{th}, u_h, u_{th}^* - u_{th}) + b(u_h, u_{th}, u_{th}^* - u_{th})|$$
$$\leq 2(\|u_{th}\|_{L^4}\|u_h\|_1\|u_{th}^* - u_{th}\|_{L^4} + \|u_{th}\|_1\|u_h\|_{L^4}\|u_{th}^* - u_{th}\|_{L^4})$$
$$\leq C(\|u_{th}\|_0^{\frac{d}{4}}\|u_{th}\|_1^{1-\frac{d}{4}}\|u_h\|_1\|u_{th}^* - u_{th}\|_0^{\frac{d}{4}}\|u_{th}^* - u_{th}\|_1^{1-\frac{d}{4}}$$
$$+ \|u_{th}\|_1\|u_h\|_0^{\frac{d}{4}}\|u_h\|_1^{1-\frac{d}{4}}\|u_{th}^* - u_{th}\|_0^{\frac{d}{4}}\|u_{th}^* - u_{th}\|_1^{1-\frac{d}{4}})$$
$$\leq C\|u_{th}\|_0^{d/2}\|u_{th}\|_1^{2-\frac{d}{2}}\|u_h\|_1 + \|u_h\|_1\|u_{th}\|_0^{\frac{d}{4}}\|u_{th}\|_1^{2-\frac{d}{4}}$$
$$= I_1 + I_2. \tag{4.68}$$

Here, using the property (2.24) of the mapping Γ_h, the inverse inequality (2.12) and the Young inequality yields

$$|I_1| \leq C\|u_{th}\|_0^{d/2}\|u_{th}\|_1^{2-\frac{d}{2}}\|u_h\|_1$$
$$\leq C\|u_{th}\|_0\|u_{th}\|_1\|u_h\|_1$$
$$\leq \frac{\mu}{4}\|u_{th}\|_1^2 + C\|u_{th}\|_0^2\|u_h\|_1^2. \tag{4.69}$$

Also, use the Young inequality twice to bind the term I_2 as follows:

$$|I_2| \leq C\|u_{th}\|_1^{3/2}\|u_{th}\|_0^{1/2}\|u_h\|_1$$
$$\leq \frac{\mu}{4}\|u_{th}\|_1^2 + C\|u_{th}\|_0^2\|u_h\|_1^4. \tag{4.70}$$

Now, combining these inequalities and (4.66) yields that

$$2|b(u_{th}, u_h, u_{th}^* - u_{th}) + b(u_h, u_{th}, u_{th}^* - u_{th})|$$
$$\leq \frac{\mu}{2}\|u_{th}\|_1^2 + C\|u_h\|_1^2(\|u_h\|_1^2 + 1)\|u_{th}\|_0^2. \tag{4.71}$$

Using all these inequalities in this lemma and Eq. (4.65) gives

$$\frac{d}{dt}\|u_{th}\|_0^2 + \mu\|u_{th}\|_1^2 + S(p_{th}, p_{th}) \leq C((1 + \|u_h\|_1^2)\|u_h\|_1^2\|u_{th}\|_0^2 + \|f_t\|_0^2). \tag{4.72}$$

Noting that $\|u_{th}(0)\|_0$ can be bounded by a constant depending on the domain, body force, and viscosity, integrating (4.78) with respect to time from 0 to s, and using Lemmas 4.1, 4.4, and 4.6 and (A4), we obtain (4.63).

In order to prove (4.64), we differentiate (4.13) in time and take $(v_h, q_h) = (u_{tth}, p_{th}) \in X_h \times M_h$ to obtain

$$\|u_{tth}\|_0^2 + \frac{\mu}{2}\frac{d}{dt}\|u_{th}\|_1^2 + b(u, u_{th}, u_{tth}^*) + b(u_{th}, u, u_{tth}^*)$$

$$+b(u_{th}, u_h - u, u_{tth}^*) + b(u_h - u, u_{th}, u_{tth}^*) \leq \frac{1}{8}\|u_{tth}\|_0^2 + C\|f_t\|_0^2. \quad (4.73)$$

Obviously, using the estimate for the trilinear term (3.10), we have

$$|b(u, u_{th}, u_{tth}^*) + b(u_{th}, u, u_{tth}^*)|$$

$$\leq C\|u\|_2\|u_{th}\|_1\|u_{tth}\|_0 \leq \frac{1}{4}\|u_{tth}\|_0^2 + C\|u\|_2^2\|u_{th}\|_1^2. \quad (4.74)$$

Applying the inverse inequality (2.12), the second inequality of (3.7), and the Hölder inequality yields

$$|b(u_{th}, u_h - u, u_{tth}^* - u_{tth}) + b(u_h - u, u_{th}, u_{tth}^* - u_{tth})|$$

$$\leq Ch^{-1}\|u_{th}\|_{L^4}\|u - u_h\|_{L^4}\|u_{tth}\|_0$$

$$\leq Ch^{-1}\|u_{th}\|_0^{\frac{d}{4}}\|u_{th}\|_1^{1-\frac{d}{4}}\|u - u_h\|_0^{\frac{d}{4}}\|u - u_h\|_1^{1-\frac{d}{4}}\|u_{tth}\|_0$$

$$\leq \frac{1}{8}\|u_{tth}\|_0^2 + Ch^{-2}\|u_{th}\|_0^{\frac{d}{2}}\|u_{th}\|_1^{2-\frac{d}{2}}\|u - u_h\|_0^{\frac{d}{2}}\|u - u_h\|_1^{2-\frac{d}{2}}. \quad (4.75)$$

As for the second term in the right-hand side in (4.75), we see that

$$h^{-2}\|u_{th}\|_0^{\frac{d}{2}}\|u_{th}\|_1^{2-\frac{d}{2}}\|u - u_h\|_0^{\frac{d}{2}}\|u - u_h\|_1^{2-\frac{d}{2}}$$

$$\leq h^{-2}\|u_{th}\|_0\|u_{th}\|_1\|u - u_h\|_0\|u - u_h\|_1 \quad (4.76)$$

with $d = 2$ or $d = 3$.

Then, combining all these estimates and (4.73), it follows from Lemmas 4.2 and 4.4 that

$$\|u_{tth}\|_0^2 + \frac{d}{dt}\left(\mu\|u_{th}\|_1^2 + S(p_{th}, p_{th})\right)$$

$$\leq C\left\{\|u\|_2^2\|u_{th}\|_1^2 + h^{-2}\|u_{th}\|_0\|u_{th}\|_1\|u - u_h\|_0\|u - u_h\|_1 + \|f_t\|_0^2\right\}. \quad (4.77)$$

Finally, multiplying by $\tau(t)$, integrating from 0 to s, and applying the Cauchy-Schwarz inequality, Lemma 4.2 and (4.63), we obtain

$$\tau(s)\left(\mu\|u_{th}(s)\|_1^2 + S(p_{th}(s), p_{th}(s))\right) + \int_0^s \tau(t)\|u_{tth}\|_0^2 dt$$

$$\leq C\left\{\|u\|_2^2 \int_0^s \tau(t)\|u_{th}\|_1^2 dt + \tau(t)h^{-2}\|u_{th}\|_0\|u - u_h\|_0 \int_0^s \|u_{th}\|_1\|u - u_h\|_1 dt\right.$$

$$\left. + \int_0^s \|f_t\|_0^2 dt\right\} + \int_0^s \left(\mu\|u_{th}\|_1^2 + S(p_{th}, p_{th})\right) dt$$

$$\leq C\left\{h^{-2}\|u_{th}\|_0\|u - u_h\|_0\left(\int_0^s \|u_{th}\|_1^2 dt\right)^{1/2}\left(\int_0^s \|u - u_h\|_1^2 dt\right)^{1/2} + \int_0^s \|f_t\|_0^2 dt\right.$$

$$\left. + \|u\|_2^2 \int_0^s \|u_{th}\|_1^2 dt\right\} + \int_0^s \left(\mu\|u_{th}\|_1^2 + S(p_{th}, p_{th})\right) dt. \tag{4.78}$$

Therefore, combining (4.78), Lemmas 4.2 and 4.6, and (A4) completes the proof of (4.64). $\square$

> **Lemma 4.11** *Under the assumptions of (A1), (A4), and Lemma 4.6, it holds that, for $s \in [0, T]$,*
>
> $$\tau\|u_t - u_{th}\|_0^2 + \mu \int_0^s \tau(t)\|e_t\|_1^2 dt \leq Ch^2. \tag{4.79}$$

Proof Differentiating (4.49) with respect to time t and noting that the L^2-projection satisfies

$$(u_{tt} - P_h u_{tt}, v_h^*) = 0,$$

we obtain

$$(P_h u_{tt} - u_{tth}, v_h^*) + \mathscr{C}_h((P_h u_t - u_{th}, J_h p_t - p_{th}); (v_h, q_h)) + b(u_t - u_{th}, u, v_h^*)$$
$$+ b(u - u_h, u_t, v_h^*) + b(u_t, u - u_h, v_h^*) + b(u, u_t - u_{th}, v_h^*)$$
$$- b(u_t - u_{th}, u - u_h, v_h^*) - b(u - u_h, u_t - u_{th}, v_h^*)$$
$$= S(p_t, q_h) - A(u_t - P_h u_t, v_h^*) - D(v_h^*, p_t - J_h p_t)$$
$$- d(u_t - P_h u_t, q_h) - S(p_t - J_h p_t, q_h). \tag{4.80}$$

Taking $(v_h, q_h) = (e_{th}, 0) = (P_h u_t - u_{th}, 0)$ in (4.80) and applying simple calculations, we see that

$$\frac{1}{2}\frac{d}{dt}\|e_{th}\|_0^2 + \mu\|e_{th}\|_1^2 + b(u_t - u_{th}, u, e_{th}^* - e_{th}) + b(u - u_h, u_t, e_{th}^* - e_{th})$$

$$+b(u_t, u - u_h, e_{th}^* - e_{th}) + b(u, u_t - u_{th}, e_{th}^* - e_{th}) - b(u_t - u_{th}, u - u_h, e_{th}^* - e_{th})$$

$$-b(u - u_h, u_t - u_{th}, e_{th}^* - e_{th}) + b(u_t - u_{th}, u, e_{th}) + b(u - u_h, u_t, e_{th})$$

$$+b(u_t, u - u_h, e_{th}) + b(u, u_t - u_{th}, e_{th}) - b(u_t - u_{th}, u - u_h, e_{th})$$

$$-b(u - u_h, u_t - u_{th}, e_{th})$$

$$= -A(u_t - P_h u_t, e_{th}^*) - D(e_{th}^*, p_t - J_h p_t). \tag{4.81}$$

By the Young inequality, it follows that

$$|b(u_t - u_{th}, u, e_{th}^* - e_{th}) + b(u, u_t - u_{th}, e_{th}^* - e_{th})|$$

$$\le C\|u_t - u_{th}\|_1\|u\|_2\|e_{th}^* - e_{th}\|_0$$

$$\le \frac{\mu}{16}\|e_{th}\|_1^2 + Ch^2(\|u_t\|_1^2 + \|u_{th}\|_1^2),$$

$$|b(u - u_h, u_t, e_{th}^* - e_{th}) + b(u_t, u - u_h, e_{th}^* - e_{th})|$$

$$\le C\|u - u_h\|_1\|u_t\|_2\|e_{th}^* - e_{th}\|_0$$

$$\le \frac{\mu}{16}\|e_{th}\|_1^2 + Ch^2\|u_t\|_2^2(\|u\|_1^2 + \|u_h\|_1^2).$$

Owing to (3.9), (3.10) and the Young inequality, we find that

$$|b(u_t - u_{th}, u, e_{th}) + b(u, u_t - u_{th}, e_{th})|$$

$$\le C\|u\|_2\|e_{th}\|_1(\|u_t - u_{th}\|_0 + \|e_{th}\|_0)$$

$$\le \frac{\mu}{16}\|e_{th}\|_1^2 + C\|u\|_2^2\|e_{th}\|_0^2 + Ch^2(\|u_t\|_2^2 + \|p_t\|_1^2),$$

$$|b(u - u_h, u_t, e_{th})| + |b(u_t, u - u_h, e_{th})|$$

$$\le C\|u_t\|_1\|e_{th}\|_1\|u - u_h\|_1$$

$$\le \frac{\mu}{16}\|e_{th}\|_1^2 + C\|u_t\|_1^2\|u - u_h\|_1^2.$$

Also, the same approach described above can be used to obtain the following estimate:

$$|b(u_t - u_{th}, u - u_h, e_{th})| + |b(u - u_h, u_t - u_{th}, e_{th})|$$

$$\le \frac{\mu}{16}\|e_{th}\|_1^2 + C(\|u_{th}\|_1^2 + \|u_t\|_1^2)\|u - u_h\|_1^2.$$

Applying (3.9), we estimate the term:

$$|b(u_t - u_{th}, u - u_h, e_{th}^* - e_{th})| \le C\|u_t - u_{th}\|_1\|u - u_h\|_1\|e_{th}\|_1$$

$$\le C(\|u_{th}\|_1 + \|u_t\|_1)\|u - u_h\|_1\|e_{th}\|_1$$

$$\le \frac{\mu}{16}\|e_{th}\|_1^2 + C(\|u_{th}\|_1^2 + \|u_t\|_1^2)\|u - u_h\|_1^2.$$

Similarly, we have

$$|b(u - u_h, u_t - u_{th}, e_{th}^* - e_{th})| \leq \frac{\mu}{16}\|e_{th}\|_1^2 + C(\|u_{th}\|_1^2 + \|u_t\|_1^2)\|u - u_h\|_1^2.$$

Then, using the same approach as for (4.26) and (4.27), we arrive at

$$|A(u_t - P_h u_t, e_{th}^*)| \leq Ch\|u_t\|_2\|e_{th}\|_1 \leq \frac{\mu}{16}\|e_{th}\|_1^2 + Ch^2\|u_t\|_2^2,$$

$$|D(e_{th}^*, p_t - J_h p_t)| \leq Ch\|p_t\|_1\|e_{th}\|_1 \leq \frac{\mu}{16}\|e_{th}\|_1^2 + Ch^2\|p_t\|_1^2.$$

Substituting these inequalities into (4.81) and integrating from 0 to s, we conclude that

$$\frac{d}{dt}\|e_{th}\|_0^2 + \mu\|e_{th}\|_1^2 \leq C\left\{h^2(\|u_t\|_2^2 + \|p_t\|_1^2 + \|u_{th}\|_1^2) + \|u\|_2^2\|e_{th}\|_0^2\right.$$

$$\left. + (\|u_t\|_1^2 + \|p_t\|_1^2 + \|u_{th}\|_1^2)\|u - u_h\|_1^2\right\}. \quad (4.82)$$

Applying these inequalities and (4.64) yields

$$\tau(s)\|e_{th}(s)\|_0^2 + \mu\int_0^s \tau(t)\|e_{th}\|_1^2 dt \leq C\left\{h^2\int_0^s \tau(\|u_t\|_2^2 + \|p_t\|_1^2 + \|u_{th}\|_1^2)dt\right.$$

$$\left. + \|u - u_h\|_1^2\int_0^s \tau(t)(\|u_t\|_1^2 + \|p_t\|_1^2 + \|u_{th}\|_0^2)dt + \int_0^s \|e_{th}\|_0^2 dt\right\},$$

which, together with (2.10), Lemmas 4.2–4.6 and 4.9, and a triangle inequality completes the proof.   □

Lemma 4.12 *Under the assumptions of (A1), (A4), and Lemma 4.6, it holds that, for $s \in [0, T]$,*

$$\tau^{1/2}(s)\|p(s) - p_h(s)\|_0 \leq Ch. \quad (4.83)$$

Proof Setting $(e_h, \eta_h) = (R_h(u, p) - u_h, L_h(u, p) - p_h)$ and using the weak coercivity (2.31) and (4.49) yield

$$\beta^*\|\eta_h\|_0 \leq \sup_{(v_h, q_h)\in X_h\times M_h} \frac{\mathscr{C}_h((e_h, \eta_h); (v_h, q_h))}{\|v_h\|_1 + \|q_h\|_0}$$

$$= \sup_{(v_h, q_h)\in X_h\times M_h}\left\{\left|\frac{(u_t - u_{th}, v_h^*) + b(u - u_h, u, v_h^* - v_h) + b(u_h, u - u_h, v_h^* - v_h)|}{\|v_h\|_1 + \|q_h\|_0}\right.\right.$$

$$\left.+ \frac{b(u - u_h, u, v_h) + b(u_h, u - u_h, v_h)}{\|v_h\|_1 + \|q_h\|_0}\right\}. \quad (4.84)$$

In view of (3.9) and the Hölder inequality, it follows that

$$|b(u - u_h, u, v_h) + b(u_h, u - u_h, v_h)| \leq C(\|u\|_1 + \|u_h\|_1)\|u - u_h\|_1 \|v_h\|_1,$$
$$|b(u_h, u - u_h, v_h^* - v_h) + b(u - u_h, u, v_h^* - v_h)| \leq C\|u\|_1 \|u - u_h\|_1 \|v_h\|_1.$$

A combination of (4.31), (4.79), and (4.84) shows that

$$\tau^{1/2}\|\eta_h\|_0 \leq C\tau^{1/2}(\|u_t - u_{th}\|_0 + \|u - u_h\|_1) \leq Ch. \tag{4.85}$$

As a result, applying (4.85) and a triangle inequality gives

$$\tau^{1/2}\|p - p_h\|_0 \leq \tau^{1/2}(\|\eta\|_0 + \|p - L_h(u, p)\|_0) \leq Ch,$$

which is (4.83).   □

Theorem 4.3 *Under the assumptions of (A1), (A4), and Lemma 4.6, it holds that, for $s \in [0, T]$ and $f \in [L^2(\Omega)]^d$,*

$$\|u - u_h\|_1 + \tau^{1/2}\|p - p_h\|_0 \leq Ch. \tag{4.86}$$

This theorem follows from Lemmas 4.9 and 4.12.

4.5 L^2-Error Estimates

Observed from the previous analysis, we can find that different analysis techniques are applied to the finite volume methods from those for the finite element methods of the nonstationary Navier–Stokes equations. As for a L^2-norm estimate for velocity, we must take special care of optimal analysis since there is only an $O(h)$ error between the test functions of the finite element methods and those of the finite volume methods.

In this section we estimate the error $\|u - u_h\|_0$ using a parabolic duality argument for a backward-in-time linearized Navier–Stokes problem [68, 71, 73]. The dual problem is to seek $(\Phi(t), \Psi(t)) \in X \times M$ such that, for $t \in [0, T]$ and $g \in L^2(0, T; Y)$,

$$(v, \Phi_t) - \mathscr{B}((v, q); (\Phi, \Psi)) - b(u, v, \Phi) - b(v, u, \Phi) = (g, u - u_h), \tag{4.87}$$

for all $(v, q) \in X \times M$, with $\Phi(T) = 0$. This problem is well-posed and has a unique solution (Φ, Ψ) satisfying [73]

$$\Phi \in C(0, T; V) \cap L^2(0, T; D(A)) \cap H^1(0, T; Y), \quad \Psi \in L^2(0, T; H^1(\Omega) \cap M).$$

We recall the following regularity result [73]:

Lemma 4.13 *Under the assumptions of (A1), the solution (Φ, Ψ) of (4.87) satisfies*

$$\sup_{0 \leq s \leq T} \|\Phi(s)\|_1^2 + \int_0^T \left(\|\Phi\|_2^2 + \|\Psi\|_1^2 + \|\Phi_t\|_0^2\right) dt \leq C \int_0^T \|u - u_h\|_0^2 \, dt.$$

$$(4.88)$$

Based on the results provided in the previous section, a duality argument is applied to overcome the lower order convergence rate of the Stokes projection defined in (4.21) by involving the Stokes projection defined in (4.10). Then, optimal analysis follows from the following two Lemmas.

Lemma 4.14 *Under the assumptions of (A1), (A4), and Lemma 4.6, it holds that, for $s \in [0, T]$,*

$$\int_0^T \|u - u_h\|_0^2 dt \leq Ch^4.$$

$$(4.89)$$

Proof Let $(\Phi_h(t), \Psi_h(t)) \in X_h \times M_h$ be the dual Galerkin projection of $(\Phi(t), \Psi(t)) \in X \times M$ such that

$$\mathcal{B}_h((v_h, q_h)); (\Phi_h, \Psi_h)) = \mathcal{B}((v_h, q_h); (\Phi, \Psi)) \quad \forall (v_h, q_h) \in X_h \times M_h (4.90)$$

Then, it holds

$$\|\Phi_h\|_1 + \|\Psi_h\|_0 \leq C(\|\Phi\|_1 + \|\Psi\|_0),$$
$$\|\Phi - \Phi_h\|_0 + h(\|\Phi - \Phi_h\|_1 + \|\Psi - \Psi_h\|_0) \leq Ch^2(\|\Phi\|_2 + \|\Psi\|_1). \quad (4.91)$$

Setting $(e, \eta) = (u - u_h, p - p_h)$ and substituting $(v_h, q_h) = (\Phi_h, \Psi_h)$ into (4.49), we find that

$$(e_t, \Phi_h^*) + \mathcal{C}_h\big((e, \eta); (\Phi_h, \Psi_h)\big) + b(u, e, \Phi_h^*)$$
$$+ b(e, u, \Phi_h^*) - b(e, e, \Phi_h^*) = S(p, \Psi_h). \quad (4.92)$$

Combining (4.92) and (4.87), taking $(v, q) = (e, \eta)$ and setting $g = e$, we see that

$$\|e\|_0^2 = \frac{d}{dt}(e, \Phi) + (e_t, \Phi_h^* - \Phi) + A(e, \Phi_h^*) - a(e, \Phi_h) + D(\Phi_h^*, \eta) + d(\Phi_h, \eta)$$
$$-a(e, \Phi - \Phi_h) - d(e, \Psi - \Psi_h) + d(\Phi - \Phi_h, \eta) + S(\eta, \Psi_h)$$
$$+b(u, e, \Phi_h^* - \Phi) + b(e, u, \Phi_h^* - \Phi) - b(e, e, \Phi_h^*) - S(p, \Phi_h).$$

That is,

$$\|e\|_0^2 = \frac{d}{dt}(e, \Phi) + (e_t, \Phi_h^* - \Phi) + A(e, \Phi_h^*) - a(e, \Phi_h) + D(\Phi_h^*, \eta) + d(\Phi_h, \eta)$$
$$-\mathscr{B}_h((e, \eta); (\Phi - \Phi_h, \Psi - \Psi_h)) + S(\eta, \Psi)$$
$$+b(u, e, \Phi_h^* - \Phi) + b(e, u, \Phi_h^* - \Phi) - b(e, e, \Phi_h^*). \tag{4.93}$$

Then, the following inequalities follows from the Hölder inequality:

$$|(e_t, \Phi - \Phi_h)| \le Ch^2(\|u_t\|_0 + \|u_{th}\|_0)\|\Phi\|_2,$$
$$|(e_t, \Phi_h - \Phi_h^*)| = |(e_t - \hat{\pi}_h e_t, \Phi_h - \Phi_h^*)|$$
$$\le Ch^2(\|u_t\|_1 + \|u_{th}\|_1)\|\Phi\|_1,$$

where $\hat{\pi}_h$ is defined in (2.43). Using the bound of the trilinear term again, we have

$$|b(e, u, \Phi_h^* - \Phi_h) + b(u, e, \Phi_h^* - \Phi_h)| \le C\|u\|_2\|e\|_1(\|\Phi - \Phi_h\|_0 + \|\Phi_h - \Phi_h^*\|_0)$$
$$\le Ch\|u\|_2\|e\|_1\|\Phi\|_1.$$

By the Hölder inequality, (2.24), and (4.91), it follows that

$$|b(e, e, \Phi_h^*)|$$
$$= |b(e, e, \Phi_h^* - \Phi_h^*) + |b(e, e, \Phi_h)|$$
$$\le C\|e\|_0^{\frac{d}{4}}\|e\|_1^{2-\frac{d}{4}}(\|\Phi_h - \Phi_h^*\|_{L^4} + \|\Phi_h\|_{L^4})$$
$$\le C\|e\|_0^{\frac{d}{4}}\|e\|_1^{2-\frac{d}{4}}\|\Phi_h\|_1$$
$$\le C\|e\|_1^2\|\Phi\|_1.$$

Also, using (4.1), (4.13) and an argument in (3.53) in Chap. 3 leads to

$$A(u, \Phi_h^*) - a(u, \Phi_h) + D(\Psi_h^*, p) + d(\Phi_h, p)$$
$$= (f - (u \cdot \nabla)u - u_t, \Phi_h^* - \Phi_h)$$
$$\le Ch^2(\|f\|_1 + \|u_t\|_1 + \|u\|_0^{1-\frac{d}{4}}\|u\|_2^{\frac{d+4}{4}} + \|u\|_1^{\frac{d}{2}}\|u\|_1^{2-\frac{d}{2}})\|\Phi\|_1.$$

Then, use (4.11), (4.90), and (4.91) to obtain

$$|\mathscr{B}_h((e, \eta); (\Phi - \Phi_h, \Psi - \Psi_h)) - S(\eta, \Psi)|$$
$$= \mathscr{B}_h((u - \bar{R}_h(u, p), p - \bar{L}_h(u, p)); (\Phi - \Phi_h, \Psi - \Psi_h)) + S(\eta, \Psi)$$
$$\le Ch^2(\|u\|_2 + \|p\|_1)(\|\Phi\|_2 + \|\Psi\|_1) + Ch\|\eta\|_0\|\Psi\|_1.$$

By a combination of these estimates, $(A4)$, and (4.90)–(4.93), we see that

$$\|e\|_0^2 \leq \frac{d}{dt}(e, \Phi) + C\Big\{ \|e\|_1^2 \|\Phi\|_1 + h\|\eta\|_0 \|\Psi\|_1$$

$$+ h^2 \Big(\|u_t\|_1 + \|u_{th}\|_1 + \|f\|_1 + \|u\|_0^{1-\frac{d}{4}} \|u\|_2^{\frac{d+4}{4}} + \|u\|_1^{\frac{d}{2}} \|u\|_2^{2-\frac{d}{2}} \Big) \|\Phi\|_2$$

$$+ h^2 (\|u\|_2 + \|p\|_1)(\|\Phi\|_2 + \|\Psi\|_1) \Big\}. \tag{4.94}$$

Integrating (4.94) from 0 to s and using the Schwarz inequality yield

$$\int_0^s \|e\|_0^2 \, dt \leq C\Big\{ h^2 \Big(\int_0^s \big(\|u_t\|_1^2 + \|u_{th}\|_1^2 + \|f\|_1 + \|u\|_2^2 \big) dt \Big)^{1/2} \cdot$$

$$\Big(\int_0^s \big(\|\Phi\|_2^2 + \|\Psi\|_1^2 \big) dt \Big)^{1/2}$$

$$+ h \Big(\int_0^s \big(\|e\|_1^2 + \|\eta\|_0^2 \big) dt \Big)^{1/2} \Big(\int_0^s \big(\|\Phi\|_2^2 + \|\Psi\|_1^2 \big) dt \Big)^{1/2}$$

$$- (e(0), \Phi(0)) + \sup_{0 \leq s \leq T} \|\Phi(s)\|_1 \int_0^s \|e\|_1^2 \, dt \Big\}. \tag{4.95}$$

In addition, by the definition of the projection P_h and the initial approximation, we have

$$|(e(0), \Phi(0))| = |(u_0 - P_h u_0, \Phi(0))| \leq Ch^2 (\|u_0\|_2 + \|p_0\|_1) \|\Phi(0)\|_1. \tag{4.96}$$

Combining $(A4)$, (4.88), (4.95) and (4.96) completes the proof of (4.89). $\qquad\square$

Theorem 4.4 *Under the assumptions of $(A1)$, $(A4)$, and Lemma 4.6, it holds that, for $s \in [0, T]$ and $f \in [H^1(\Omega)]^d$,*

$$\|u - u_h\|_0 \leq C\tau^{-1/2} h^2. \tag{4.97}$$

Proof Taking $(v_h, q_h) = (\bar{e}_h, \bar{\eta}_h) = (\bar{R}_h(u, p) - u_h, \bar{L}_h(u, p) - p_h)$ and setting $\bar{E} = u - \bar{R}_h(u, p)$ in (4.49), we see that

$$\frac{1}{2}\frac{d}{dt}\|\bar{e}_h\|_0^2 + \mu \|\bar{e}_h\|_1^2 + S(\bar{\eta}_h, \bar{\eta}_h) + b(u, u - u_h, \bar{e}_h^* - \bar{e}_h) + b(u - u_h, u, \bar{e}_h^* - \bar{e}_h)$$

$$- b(u - u_h, u - u_h, \bar{e}_h^* - \bar{e}_h) + b(u, u - u_h, \bar{e}_h) + b(u - u_h, u, \bar{e}_h)$$

$$- b(u - u_h, u - u_h, \bar{e}_h)$$

$$= S(p, \bar{e}_h^*) - (\bar{E}_t, \bar{e}_h^*) - A(\bar{E}, \bar{e}_h^*) - D(\bar{e}_h^*, p - \bar{L}_h(u, p)) + d(u - L_h(u, p), \eta_h). \tag{4.98}$$

Clearly, using Lemmas 2.2 and the Young inequality, we find that

$$|(E_t, \bar{e}^*)| \leq \|E_t\|_0 \|\bar{e}\|_0 \leq Ch^2(\|u_t\|_2 + \|p_t\|_1)\|\bar{e}\|_0$$
$$\leq Ch^4(\|u_t\|_2^2 + \|p_t\|_1^2) + \|\bar{e}\|_0^2.$$

Using the estimates of the trilinear term again gives

$$|b(u, u - u_h, \bar{e}) + b(u - u_h, u, \bar{e})| \leq C\|u\|_2 \|\bar{e}\|_1 \|u - u_h\|_0$$
$$\leq \frac{\mu}{10}\|\bar{e}\|_1^2 + C\|u\|_2^2\|u - u_h\|_0^2,$$
$$|b(u - u_h, u - u_h, \bar{e}) + b(u - u_h, u - u_h, \bar{e}^* - \bar{e})| \leq \|u - u_h\|_1^2 \|\bar{e}\|_1$$
$$\leq \frac{\mu}{10}\|\bar{e}\|_1^2 + C\|u - u_h\|_1^4.$$

By the technique of the trilinear terms, it follows that

$$|b(u, u - u_h, \bar{e}^* - \bar{e}) + b(u - u_h, u, \bar{e}^* - \bar{e})| \leq C\|u\|_2 \|u - u_h\|_1 \|\bar{e}^* - \bar{e}\|_0$$
$$\leq Ch^2\|u\|_2^2\|u - u_h\|_1^2 + \frac{\mu}{10}\|\bar{e}\|_1^2.$$

Also, it follows from the definition of $\hat{\pi}_h$, an argument in (3.53) and the Young inequality that

$$|A(\bar{E}, \bar{e}_h^*) + D(\bar{e}_h^*, p - \bar{L}_h(u, p))|$$
$$= |([f - \hat{\pi}_h f] - [(u \cdot \nabla)u - \hat{\pi}_h(u \cdot \nabla)u] - [u_t - \hat{\pi}_h u_t], \bar{e}_h^* - \bar{e}_h)|$$
$$\leq Ch^2(\|f\|_1 + \|u_t\|_1 + \|u\|_0^{1-\frac{d}{4}}\|u\|_2^{\frac{d+4}{4}} + \|u\|_1^{\frac{d}{2}}\|u\|_2^{2-\frac{d}{2}})\|\bar{e}_h\|_1$$
$$\leq \frac{\mu}{10}\|\bar{e}_h\|_1^2 + Ch^4(\|f\|_1^2 + \|u\|_2^2 + \|u_t\|_1^2).$$

Therefore, combining these estimates and (4.4) gives

$$\frac{d}{dt}\|\bar{e}_h\|_0^2 + \mu\|\bar{e}_h\|_1^2 + S(\bar{\eta}_h, \bar{\eta}_h)$$
$$\leq C\left\{ h^4(\|u_t\|_2^2 + \|p_t\|_1^2 + \|f\|_1^2 + \|u\|_2^2) + \|\bar{e}_h\|_0^2 + h^2\|u\|_2^2\|u - u_h\|_1^2 \right.$$
$$\left. + \|u - u_h\|_1^4 + h^2\|u - u_h\|_1^2 + \|u\|_2^2\|u - u_h\|_0^2 \right\}. \tag{4.99}$$

Using Lemmas 4.2 and 4.3, we have

$$\|\bar{E}\|_0^2 \leq Ch^4(\|u\|_2^2 + \|p\|_1^2) \leq Ch^4. \tag{4.100}$$

Noting that

$$
\int_0^s \|\bar{e}_h\|_0^2 \, dt \leq \int_0^s \|u - u_h\|_0^2 \, dt + \int_0^s \|\bar{E}\|_0^2 \, dt
$$
$$
\leq \int_0^s \|u - u_h\|_0^2 \, dt + Ch^4 \int_0^s (\|u\|_2^2 + \|p\|_1^2) \, dt \leq Ch^4, \quad (4.101)
$$

multiplying (4.99) by $\tau(t)$, integrating from 0 to s, and using (4.99)–(4.101) and Lemmas 4.2 and 4.4, we obtain

$$
\tau(s)\|\bar{e}_h(s)\|_0^2 + \int_0^s \tau(t) \left(\mu \|\bar{e}_h\|_1^2 + S(\eta, \eta) \right) dt
$$
$$
\leq C \left\{ \int_0^s \|\bar{e}_h\|_0^2 dt + h^4 \int_0^t \tau(t)(\|u\|_2^2 + \|u_t\|_2^2 + \|p_t\|_1^2 + \|f\|_1^2) dt \right.
$$
$$
\left. + \int_0^s \|u - u_h\|_1^4 dt + h^2 \int_0^s \|u - u_h\|_1^2 dt \right\}
$$
$$
\leq Ch^4, \quad (4.102)
$$

which, together with (4.100), yields (4.97). □

4.6 Conclusions

In this chapter, we analyze the stabilized finite volume methods for the nonstationary Navier–Stokes equations based on a relationship between the finite element methods and the finite volume methods and some additional analytical techniques. As noted earlier, there are several difficulties in analyzing these finite volume methods for the nonstationary Navier–Stokes equations. Some remarks need be made. First, the analysis requires a slightly extra regularity on the source force to obtain an optimal estimate in the L^2-norm for velocity. Second, additional attention is here required to treat the trilinear term for the nonlinear Naiver-Stokes equations because of its losing the anti-symmetric property. Third, additional techniques need be provided for the parabolic system that is in the form of the Petrov–Galerkin system generated by two different Stokes projections introduced above.

In this chapter, we strictly prove and adjust some lemmas and theories in accordance with the framework of the results introduced in [85, 93]. From the point of view of implementation, we analyze the discrete methods in time, with suitably small data, and uniqueness of a suitably small solution, without smooth assumptions (the data f is small and/or the viscosity is large). Therefore, the approach presented here is fairly robust and adapts to the important case of the free flow that is not too fast in practice.

References

1. R.A. Adams, *Sobolev Spaces* (Academic, New York, 1975)
2. M. Ainsworth, J.T. Oden, *A Posteriori Error Estimation in Finite Element Analysis*. Pure and Applied Mathematics (Wiley-Interscience, Wiley, New York, 2000)
3. D.N. Arnold, F. Brezzi, M. Fortin, A stable finite element for the Stokes equations. Calcolo **21**, 337–344 (1984)
4. D.N. Arnold, L.R. Scott, M. Vogelius, Regular inversion of the divergence operator with dirichlet boundary conditions on a polygon. Ann. Scuola. Norm. Sup. Pisa Cl. Sci.-serie **19**, 162–192 (1988)
5. M. Arroyo, M. Ortiz, Local maximum-entropy approximation schemes: a seamless bridge between finite elements and meshfree methods. Int. J. Numer. Methods Eng. **65**, 2167–2202 (2006)
6. I. Babuska, U. Banerjee, J.E. Osborn, Survey of meshless and generalized finite element methods: a unified approach. Acta Numer. **12**, 1–125 (2003)
7. R.E. Bank, D.J. Rose, Some error estimates for the box method. SIAM J. Numer. Anal. **24**, 777–787 (1987)
8. E. Bänsch, P. Morin, R.H. Nochetto, An adaptive uzawa fem for the Stokes problem convergence without the inf sup condition. Numer. Math. **40**, 1207–1229 (2002)
9. P. Binev, W. Dahmen, R. Devore, Adaptive finite element methods with convergence rates. Numer. Math. **97**, 219–268 (2004)
10. J.E. Bishop, A displacement-based finite element formulation for general polyhedra using harmonic shape functions. Int. J. Numer. Methods Eng. **97**, 1–31 (2014)
11. P. Bochev, C.R. Dohrmann, M.D. Gunzburger, Stabilization of low-order mixed finite elements for the Stokes equations. SIAM J. Numer. Anal. **44**, 82–101 (2006)
12. P.B. Bochev, J.M. Hyman, Principles of mimetic discretizations of differential operators, in *Compatible Spatial Discretizations*. The IMA Volumes Mathematics and Its Application, vol. 142 (2006), pp. 89–119
13. S. Brenner, L.R. Scott, *The Mathematical Theory of Finite Elements* (Springer, New York, 1991)
14. F. Brezzi, J. Douglas Jr., Stabilized mixed methods for the Stokes problem. Numer. Math. **53**, 225–235 (1988)
15. F. Brezzi, M. Fortin, *Mixed and Hybrid Finite Element Methods* (Springer, New York, 1991)
16. F. Brezzi, J. Rappaz, P.A. Raviart, Finite-dimensional approximation of nonlinear problems. Part I: branches of nonsingular solutions. Numer. Math. **36**, 1–25 (1980)

© The Author(s), under exclusive license to Springer Nature Switzerland AG 2022 115
J. Li et al., *Finite Volume Methods for the Incompressible Navier–Stokes Equations*,
SpringerBriefs in Mathematical Methods,
https://doi.org/10.1007/978-3-030-94636-4

17. E. Burman, Pressure projection stabilizations for Galerkin approximations of Stokes' and Darcy's problem. Numer. Methods Partial Diff. Equ. **24**, 285–311 (2008)
18. G.C. Buscagliaa, F.G. Basombrio, R. Codinab, Fourier analysis of an equal-order incompressible flow solver stabilized by pressure gradient projection. Int. J. Numer. Method Fluids **34**, 65–92 (2000)
19. A. Cangiani, G. Manzini, A. Russo, Convergence analysis of the mimetic finite difference method for elliptic problems. SIAM J. Numer. Anal. **47**, 2612–2637 (2009)
20. Z. Cai, J. Mandel, S. McCormick, The finite volume element method for diffusion equations on general triangulations. SIAM J. Numer. Anal. **28**, 392–403 (1991)
21. C. Cancès, I. Pop, M. Vohralík, An a posteriori error estimate for vertex-centered finite volume discretizations of immiscible incompressible two-phase flow. Math. Comput. **83**, 153–188 (2014)
22. C. Carstensen, R. Lazarov, S. Tomov, Explcit and averaging a posteriori error estimates for adaptive finite volume methods. SIAM J. Numer. Anal. **42**, 2496–2521 (2005)
23. C. Carstensen, R.H.W. Hoppe, Convergence analysis of an adaptive nonconforming finite element method. Numer. Math. **103**, 251–266 (2006)
24. N. Chalhoub, A. Ern, T. Sayah, M. Vohralík, A posteriori error estimates for unsteady convection–diffusion–reaction problems and the finite volume method, in *Finite Volumes for Complex Applications VI Problems and Perspectives* (Springer, Berlin, Heidelberg, 2011), pp. 215–223
25. L. Chen, *Navier-Stokes Equations for Fluid Dynamics*. Lecture Notes (2014)
26. Z. Chen, *Finite Element Methods and Their Applications* (Spring, Heidelberg, 2005)
27. Z. Chen, On the control volume finite element methods and their applications to multiphase flow. Netw. Heterog. Media **1**, 689–706 (2006)
28. Z. Chen, R. Li, A. Zhou, A note on the optimal L^2-estimate of finite volume element method. Adv. Comput. Math. **16**, 291–303 (2002)
29. E. Chénier, R. Eymard, T. Gallouët, R. Herbin, An extension of the MAC scheme to locally refined meshes: convergence analysis for the full tensor time-dependent Navier–Stokes equations. Calcolo **52**, 69–107 (2015)
30. A. Chorin, J. Marsden, *A Mathematical Introduction to Fluid Mechanics* (Springer, Berlin, 1993)
31. S.H. Chou, D.Y. Kwak, Analysis and convergence of a MAC scheme for the generalized Stokes problem. Numer. Meth. Partial Diff. Equ. **13**, 147–162 (1997)
32. S.H. Chou, D.Y. Kwak, A covolume method based on rotated bilinears for the generalized Stokes problem. SIAM J. Numer. Anal. **35**, 494–507 (1998)
33. S.H. Chou, D.Y. Kwak, Mixed covolume methods on rectangular grids for elliptic problems. SIAM J. Numer. Anal. **37**, 758–771 (2000)
34. S.H. Chou, Q. Li, Error estimates in L^2, H^1 and L^∞ in co-volume methods for elliptic and parabolic problems: a unified approach. Math. Comput. **69**, 103–120 (2000)
35. S.H. Chou, P.S. Vassilevski, A general mixed co-volume framework for constructing conservative schemes for elliptic problems. Math. Comput. **68**, 991–1011 (1999)
36. S.H. Chou, D.Y. Kwak, K.Y. Kim, A general framework for constructing and analyzing mixed finite volume methods on quadrilateral grids: the overlapping covolume case. SIAM J. Numer. Anal. **39**, 1170–1196 (2001)
37. P.G. Ciarlet, *The Finite Element Method for Elliptic Problems* (North-Holland, Amsterdam, 1978)
38. P. Clement, Approximation by finite element functions using local regularization. Math. Comput. RAIRO Anal. Numer. **9**, 77–84 (1975)
39. B. Cockburn, J. Gopalakrishnan, R. Lazarov, Unified hybridization of discontinuous Galerkin, mixed, and continuous Galerkin methods for second-order elliptic problems. SIAM J. Numer. Anal. **47**, 1319–1365 (2009)
40. Y. Coudière, G. Manzini, The discrete duality finite volume method for convection-diffusion problems. SIAM J. Numer. Anal. **47**, 4163–4192 (2010)

41. S. Delcourte, K. Domelevo, P. Omnes, A discrete duality finite volume approach to hodge decomposition and div-curl problems on almost arbitrary two-dimensional meshes. SIAM J. Numer. Anal. **45**, 1142–1174 (2007)
42. J. Douglas Jr., J. Wang, An absolutely stabilized finite element method for the Stokes problem. Math. Comput. **52**, 495–508 (1989)
43. W. Dörfler, A convergent adaptive algorithm for Poisson equation. SIAM J. Numer. Anal. **33**, 1106–1124 (1996)
44. J. Droniou, R. Eymard, T. Gallouët, R. Herbin, A unified approach to mimetic finite difference, hybrid finite volume and mixed finite volume methods. Math. Models Methods Appl. Sci. **20**, 265–295 (2010)
45. J. Droniou, R. Eymard, T. Gallouët, R. Herbin, Gradient schemes: a generic framework for the discretisation of linear, nonlinear and nonlocal elliptic and parabolic equations. Math. Models Methods Appl. Sci. **23**, 1142–1174 (2013)
46. R. Duran, R. Nochetto, J. Wang, Sharp Maximum norm error estimates for finite element approximations of the Stokes problem in 2-D. Math. Comput. **15**, 491–506 (1988)
47. K. Eriksson, D. Estep, P. Hansbo, C. Johnson, Introduction to adaptive methods for differential equations, in *Acta Numerica 1995*, ed. by A. Iserles (Cambridge University Press, Cambridge, 1995), pp. 105–58
48. A. Ern, I. Smears, M. Vohralík, Equilibrated flux a posteriori error estimates in $L^2(H^1)$-norms for high-order discretizations of parabolic problems. arXiv:1703.04987 [math.NA]
49. R.E. Ewing, T. Lin, Y. Lin, On the accuracy of the finite volume element method based on piecewise linear polynomials. SIAM J. Numer. Anal. **39**, 1865–1888 (2002)
50. R. Eymard, J. Fuhrmann, A. Linke, On MAC schemes on triangular delaunay meshes, their convergence and application to coupled flow problems. Numer. Meth. Partial Diff. Equ. **30**, 1397–1424 (2014)
51. R. Eymard, T. Gallouët, R. Herbin, J.C. Latché, Convergence of the MAC scheme for the compressible Stokes equations. Numer. Meth. Partial Diff. Equ. **56**, 941–950 (2017)
52. R. Eymard, C. Guichard, R. Herbin, R. Masson, Gradient schemes for two-phase flow in heterogeneous porous media and Richards equation. ZAMM - J. Appl. Math. Mech. **94**, 560–585 (2014)
53. R. Eymard, R. Herbin, J.C. Latché, B. Piar, Convergence analysis of a locally stabilized collocated finite volume scheme for incompressible flows. ESAIM: M2AN **43**, 889–927 (2009)
54. M. Floater, A. Gillette, N. Sukumar, Gradient bounds for Wachspress coordinates on polytopes. SIAM J. Numer. Anal. **52**, 515–532 (2014)
55. T.P. Fries, T. Belytschko, The extended/generalized finite element method: an overview of the method and its applications. Int. J. Numer. Methods Eng. **84**, 253–304 (2010)
56. T. Gallouët, R. Herbin, J.C. Latché, K. Mallem, Convergence of the marker-and-cell scheme for the incompressible Navier-Stokes equations on non-uniform grids. Found. Comput. Math. **17**, 1–41 (2016)
57. V. Girault, R.H. Nochetto, R. Scott, Maximum-norm stability of the finite element Stokes projection. J. Math. Pures Appl. **84**, 279–330 (2005)
58. V. Girault, R.H. Nochetto, R. Scott, Max-norm estimates for Stokes and Navier-Stokes approximations in convex polyhedra. Numer. Math. **131**, 771–822 (2015)
59. V. Girault, P.A. Raviart, *Finite Element Method for Navier-Stokes Equations: Theory and Algorithms* (Springer, Berlin, 1987)
60. C.C. Hillairet, J. Droniou, Convergence analysis of a mixed finite volume scheme for an elliptic-parabolic system modeling miscible fluid flows in porous media. SIAM J. Numer. Anal. **45**, 2228–2258 (2007)
61. G. Kanschat, Divergence-free discontinuous Galerkin schemes for the Stokes equations and the MAC scheme. Numer. Meth. Partial Diff. Equ. **56**, 941–950 (2017)
62. S. Krell, G. Manzini, The discrete duality finite volume method for Stokes equations on three-dimensional polyhedral meshes. SIAM J. Numer. Anal. **50**, 808–837 (2012)

63. D.Y. Kwak, A new class of higher order mixed finite volume methods for elliptic problems. Math. Models Methods Appl. Sci. **50**, 1941–1958 (2012)
64. R.A. Nicolaides, Analysis and convergence of the MAC scheme. I. The linear problem. SIAM J. Numer. Anal. **29**, 1579–1591 (1992)
65. P. Omnes, Y. Penel, Y. Rosenbaum, A posteriori error estimation for the discrete duality finite volume discretization of the Laplace equation. SIAM J. Numer. Anal. **47**, 2782–2807 (2009)
66. B. Ödonoghue, E. Candès, Adaptive restart for accelerated gradient schemes, SIAM J. Numer. Anal. **15**, 715–732 (2015)
67. Y. He, J. Li, A stabilized finite element method based on local polynomial pressure projection for the stationary Navier-Stokes equation. Appl. Numer. Math. **58**, 1503–1514 (2008)
68. Y. He, Y. Lin, W. Sun, Stabilized finite element method for the non-stationary Navier-Stokes problem. Discret. Contin. Dyn. Syst.-Ser. B **6**, 41–68 (2006)
69. Y. He, A. Wang, L. Mei, Stablized finite-element method for the stationary Navier-Stokes equations. J. Eng. Math. **51**, 367–380 (2005)
70. Y. He, W. Sun, Stabilized finite element methods based on Crank-Nicolson extrapolation scheme for the time-dependent Navier-Stokes equations. Math. Comput. **76**, 115–136 (2007)
71. J.G. Heywood, R. Rannacher, Finite element approximation of the nonstationary Navier-Stokes problem I: regularity of solutions and second-order error estimates for spatial discretization. SIAM J. Numer. Anal. **19**, 275–311 (1982)
72. J.G. Heywood, R. Rannacher, Finite element approximation of the nonstationary Navier-Stokes problem III: Smothing property and higher order error estimates for spatial discretization. SIAM J. Numer. Anal. **25**, 489–582 (1988)
73. A.T. Hill, E. Süli, Approximation of the global attractor for the incompressible Navier-Stokes problem. IMA J. Numer. Anal. **20**, 633–667 (2000)
74. J. Huang, S. Xi, On the finite volume element method for general self-adjoint elliptic problems. SIAM J. Numer. Anal. **35**, 1762–1774 (1998)
75. T. Hughes, L. Franca, M. Balestra, A new finite element formulation for computational fluid dynamics: V. Circumventing the Babuska-Brezzi condition: a stable Petrov-Galerkin formulation of the Stokes problem accommodating equal-order interpolations. Comput. Methods Appl. Mech. Eng. **59**, 85–99 (1986)
76. D. Kay, D. Silvester, A posteriori error estimation for stabilized mixed approximations of the Stokes equations. SIAM J. Sci. Comput. **21**, 1321–1337 (2000)
77. N. Kechkar, D. Silvester, Analysis of locally stabilized mixed finite element methods for the Stokes problem. Math. Comput. **58**, 1–10 (1992)
78. Y.J. Kim, *A Mathematical Introduction to Fluid Mechanics*. Lecture Notes (2008)
79. Y. Kondratyuk, R. Stevenson, An optimal adaptive finite element method for the Stokes problem. SIAM J. Numer. Anal. **46**, 747–775 (2008)
80. O.A. Ladyzhenskaya, *The Mathematical Theory of Viscous Incompressible Flow*. Mathematics and Its Applications (Gordon and Breach Science Publishers, London, 1969)
81. W. Layton, Model reduction by constraints, discretization of flow problems and an induced pressure stabilization. Numer. Linear Algebra Appl. **12**, 547–562 (2005)
82. R. Li, On generalized difference methods for elliptic and parabolic differential equations, in *Proceeding of the Symposium on the Finite Element Method between China and France*, ed. by K. Feng, J.L. Lions (Science Press, Beijing, China, 1982), pp. 323–360
83. R. Li, Generalized difference methods for a nonlinear Dirichlet problem. SIAM J. Numer. Anal. **24**, 77–88 (1987)
84. J. Li, Z. Chen, A new stabilized finite volume method for the stationary Stokes equations. Adv. Comput. Math. **30**, 141–152 (2009)
85. J. Li, Z. Chen, On the semi-discrete stabilized finite volume method for the transient Navier-Stokes equations. Adv. Comput. Math. **38**, 281–320 (2013)
86. J. Li, Z. Chen, Optimal L^2, H^1 and L^∞ analysis of finite volume methods for the stationary Navier-Stokes equations with large data. Numer. Math. **126**, 75–101 (2014)
87. J. Li, Z. Chen, Optimal L^∞ analysis of non-singular finite element methods/finite volume methods for the stationary 3D Navier-Stokes equations. Sci. China Math. (in Chinese) **45**, 1281–1298 (2015)

88. J. Li, Z. Chen, Y. He, A stabilized multi-level method for non-singular finite volume solutions of the stationary 3D Navier-Stokes equations. Numer. Math. **122**, 279–304 (2012)
89. J. Li, Y. He, the property of the branch of nonsingular finite element/finite volume solutions to the stationary Navier-Stokes equations and its application. Int. J. Numer. Anal. Model. **17**, 42–53 (2020)
90. R. Li, Z. Chen, W. Wu, *The Generalized Difference Method for Differential Equations-numerical Analysis of Finite Volume Methods* (Marcel Dekker, New York, 2000)
91. J. Li, Y. He, A stabilized finite element method based on two local gauss integrations for the Stokes equations. J. Comp. Appl Math. **214**, 58–65 (2008)
92. J. Li, Y. He, Z. Chen, A new stabilized finite element method for the transient Navier-Stokes equations. Comp. Meth. Appl. Mech. Eng. **197**, 22–35 (2007)
93. J. Li, X. Lin, X. Zhao, Optimal estimates on stabilized finite volume methods for the incompressible Navier-Stokes model in three dimensions. Numer. Meth. Part. D. E. **35**, 128–154 (2019)
94. J. Li, Y. He, Z. Chen, Performance of several stabilized finite element methods for the Stokes equations based on the lowest equal-order pairs. Computing **86**, 37–51 (2009)
95. J. Li, L. Shen, Z. Chen, Convergence and stability of a stabilized finite volume method for the stationary Navier-Stokes equations. BIT Numer. Math. **50**, 823–842 (2010)
96. J. Li, X. Zhao, J. Wu, Study of stabilization of the lower order finite volume methods for the incompressible flows. Acta Math. Sinica **56** (2013) (In Chinese)
97. J. Li, X. Zhao, Z. Chen, A novel l^{∞} analysis for finite volume approximations of the Stokes problem. J. Comp. Appl Math. **279**, 97–105 (2015)
98. R. Li, P. Zhu, Generalized difference methods for second order elliptic partial differential equations (I) (in Chinese). Numer. Math. J. Chin. U. **4**, 140–152 (1982)
99. X. Li, J. Li, Z. Chen, A nonconforming virtual element method for the Stokes problem on general meshes. Comput. Methods Appl. Mech. Eng. **320**, 694–711 (2017)
100. J. Lv, Y. Li, L^2 error estimate of the finite volume element methods on quadrilateral meshes. Adv. Comput. Math. **33**, 129–148 (2010)
101. J. Lv, Y. Li, L^2 error estimates and superconvergence of the finite volume element methods on quadrilateral meshes. Adv. Comput. Math. **37**, 393–416 (2012)
102. P. Markowich, *Applied Partial Differential Equations: A Visual Approach* (Springer, Berlin, 2007)
103. P. Morin, R.h. Nochetto, K.G. Siebert, Data oscillation and convergence of adaptive FEM. SIAM J. Numer. Anal. **2**, 466–488 (2000)
104. P. Morin, R.h. Nochetto, K.G. Siebert, Local problems on stars: a posteriori error estimators, convergence, and performance. Math. Comp. **72**, 1067–1097 (2002)
105. P. Morin, R.H. Nochetto, K.G. Siebert, Convergence of adaptive finite element methods. SIAM Rev. **44**, 631–658 (2002)
106. P. Morin, K.G. Siebert, A. Veeser, A basic convergence result for conforming adaptive finite element methods. Math. Models Methods Appl. Sci. **18**, 707–737 (2008)
107. S. Norburn, D. Silvester, Stabilised vs stable mixed methods for incompressible flow. Comput. Methods Appl. Mech. Eng. **166**, 1–10 (1998)
108. A. Quarteroni, *Numerical Models for Differential Problems* (Springer, Berlin, 2009)
109. A. Quarteroni, R. Sacco, F. Saleri, *Numerical Mathematics*, 2nd edn. (Springer, Berlin, 2007)
110. R. Rannacher, R. Scott, Some optimal error estimates for piecewise linear finite element approximations. Math. Comput. **38**, 437–445 (1982)
111. T. Rabczuk, S. Bordas, G. Zi, On three-dimensional modelling of crack growth using partition of unity methods. Comput. Struct. **88**, 1391–1411 (2010)
112. A. Rand, A. Gillette, C. Bajaj, Interpolation error estimates for mean value coordinates over convex polygons. Adv. Comput. Math. **39**, 327–347 (2013)
113. S. Rjasanow, S. Weisser, FEM with trefftz trial functions on polyhedral elements. J. Comput. Appl. Math. **263**, 202–217 (2014)
114. H. Rui, Analysis on a finite volume element method for the Stokes problems. Acta Math. Appl. Sin. English Series **3**, 359–372 (2005)

115. L.R. Scott, S. Zhang, Finite element interpolation of nonsmooth functions satisfying boundary conditions. Math. Comput. **54**, 483–493 (1990)
116. J. Shen, On error estimates of the penalty method for unsteady Navier-Stokes equations. SIAM J. Numer. Anal. **32**, 386–403 (1995)
117. L. Shen, J. Li, Z. Chen, Analysis of a stabilized finite volume method for the transient Stokes equations. Int. J. Numer. Anal. Mod. **6**, 505–519 (2009)
118. D. Silvester, Stabilized mixed finite element methods, Numerical Analysis Report, No. 262 (1995)
119. D. Silvester, N. Kechkar, Stabilised bilinear-constant velocity-pressure finite elements for the conjugate gradient solution of the Stokes problem. Comput. Methods Appl. Mech. Eng. **79**, 71–86 (1990)
120. R. Stenberg, Analysis of mixed finite element methods for the Stokes problem: a unified approach. Math. Comput. **42**, 9–23 (1984)
121. R. Steveneon, Optimality of a standard adaptive finite element method. Found. Comput. Math. **7**, 245–269 (2007)
122. R. Temam, *Navier-Stokes Equations* (North-Holland, Amsterdam, 1984)
123. V. Thomée, *Galerkin Finite Element Methods for Parabolic Problems*, Lecture Notes in Math, vol. 1054 (Springer, Berlin, 1984)
124. L.B.d. Veiga, G. Manzini, A higher-order formulation of the mimetic finite difference method. SIAM J. Sci. Comput. **31**, 732–760 (2008)
125. L.B.d. Veiga, K. Lipnikov, G. Manzini, *The Mimetic Finite Difference Method for Elliptic Problems* (Springer International Publishing, Switzerland, 2014)
126. R. Verfürth, A posteriori error estimators for the Stokes equations. Numer. Math. **55**, 309–325 (1989)
127. R. Verfürth, *A Review of a Posteriori Error Estimation and Adaptive Mesh-Refinement Techniques* (Wiley-Teubner, Chichester, 1996)
128. M. Vohralík, Residual flux-based a posteriori error estimates for finite volume and related locally conservative methods. Numer. Math. **111**, 121–158 (2008)
129. M. Vohralík, Residual flux-based a posteriori error estimates for finite volume discretizations of inhomogeneous, anisotropic, and convection-dominated problems (2006). <hal-00079221v1>
130. J. Wang, Y. Wang, X. Ye, A new finite volume method for the stokes problems. Int. J. Numer. Anal. Mod. **7**, 281–302 (2009)
131. H. Wu, R. Li, Error estimates for finite volume element methods for general second-order elliptic problems. Numer. Meth. Partial Diff. Equ. **19**, 693–708 (2003)
132. J. Xu, Y. Zhu, Q. Zou, New adaptive finite volume methods and convergence analysis, preprint (2010)
133. J. Xu, Q. Zou, Analysis of linear and quadratic simplicial finite volume methods for elliptic equations. Numer. Math. **111**, 469–492 (2009)
134. X. Ye, On the relationship between finite volume and finite element methods applied to the Stokes equations. Numer. Methods Partial Diff. Equ. **5**, 440–453 (2001)
135. T. Zhang, L. Tang, A stabilized finite volume method for Stokes equations using the lowest order $P_1 - P_0$ element pair. Adv. Comp. Math. **41**, 781–798 (2015)
136. Z. Zhang, Q. Zou, Some recent advances on vertex centered finite volume element methods for elliptic equations. Sci. China Math. **56**, 2507–2522 (2013)
137. https://en.wikipedia.org/wiki/

Index

A
Adaptive, 9, 11, 32, 38, 42–44, 51, 52

B
Banach spaces, 1, 2

C
Convergence analysis, 25, 39, 43, 44, 61, 63, 83, 85

E
Existence and uniqueness theorems, 19, 54

F
Finite Element Method (FEM), 9–11, 14, 15, 22–25, 30, 38–40, 43, 51, 53, 57, 71, 75, 78, 79, 82, 83, 88, 89, 93, 108, 113
Finite Volume Method (FVM) , vii, 9–11, 16–18, 23, 32, 39, 40, 42–44, 47, 51–53, 57, 61, 66, 71, 76–78, 80, 82, 83, 85, 86, 89, 92, 93, 108, 113

H
H^1-norm, 10, 51, 63, 81, 82

L
L^2-norm, 9, 10, 12, 24, 51, 53, 54, 63, 64, 81–83, 85, 90, 108, 113
L^∞-norm, 25, 27, 31, 51, 53, 54, 66, 81

Lower bound, 11, 32, 35, 37, 39–42, 46, 52
Lower order finite element, vii, 25, 32, 39, 50, 51, 78
Lower order finite volume, 53, 85
L^p space, 1, 2

N
Navier–Stokes equations, vii, 4–6, 31, 53, 54, 56, 57, 61, 63, 66, 70, 74–78, 80–83, 85, 86, 89, 93, 108, 113
Nonsingular solutions, 53, 66, 74, 76–79, 81, 82
Numerical analysis, vii, 54, 87
Numerical experiments, 47, 81

O
Optimal analysis, vii, 23, 31, 63, 77, 80, 81, 83, 108, 109

P
Pressure, 5, 9–11, 15, 25, 31, 39, 47–51, 53–55, 63, 68, 72, 73, 81–83, 86
Priori estimate, 11, 22

S
Small data, 53, 54, 82
Sobolev spaces, 1, 2, 12
Stability, 10, 11, 14, 24, 25, 54, 57, 66–68, 72, 75, 77, 79, 82, 83, 86, 91–93
Stabilized finite element methods, 51, 52, 78, 85

Stabilized finite volume methods, 11, 16, 17,
        22, 32, 47, 50, 52, 81, 82, 85, 86, 93,
        113
Superclose, 22, 25, 51, 53, 61, 63, 79, 81

U
Upper bound, 32, 51

V
Velocity, 5, 6, 9–11, 24, 25, 27, 31, 39, 47–
        51, 53–55, 63, 64, 68, 72, 73, 81–83,
        85–87, 108, 113

W
Weak formulation, 11, 12, 54, 55, 75, 86

Printed in the United States
by Baker & Taylor Publisher Services